FIGHTING Light Pollution

Smart Lighting Solutions
for Individuals and
Communities

The International Dark-Sky Association

STACKPOLE
BOOKS

0 11557 03637 4

Published by
STACKPOLE BOOKS
5067 Ritter Road
Mechanicsburg, PA 17055
www.stackpolebooks.com

Printed in China

10 9 8 7 6 5 4 3 2 1

First edition

Cover photo of El Paso skyline by Vallarie Enriquez
Cover photo of night sky by Nadezhda Bolotina
Cover design by Tessa Sweigert

Library of Congress Cataloging-in-Publication Data

Fighting light pollution : smart lighting solutions for individuals and communities / the International Dark-Sky Association. — 1st ed.
    p. cm.
  Includes bibliographical references.
  ISBN 978-0-8117-3637-4
  1. Light pollution. 2. Exterior lighting. I. International Dark-Sky Association.
  QB51.3.L53F54 2012
  621.32'290286—dc23
                              2011033343

Contents

Our Endangered Night

In 1609, Galileo Galilei trained a telescope on the night sky. A year later, he published his observations in a slim booklet he titled *Sidereus Nuncius,* or *Starry Messenger.* His inspection of the paths of stars and planets brought profound new knowledge of celestial movement, and his discovery that the earth is not the center of the universe propelled revolutions in science and religion.

Remarkable as they were, Galileo's telescopic discoveries—and the cultural commotions they caused—were less spectacular than Galileo's night sky. Four hundred years ago, the twinkling stars cast shadows on a moonless night. Constellations were so intricate as to be unrecognizable today, and some sky landmarks were defined not by individual stars but by the dark spaces between them. The Milky Way appeared so solid that only a magnified view revealed it wasn't one body but hundreds of thousands of stars so densely clustered that their faint, ancient light formed a path across the cosmos.

It's hard to imagine it today, but the nocturnal world was once a central part of human existence. The night has beguiled mankind since prehistoric days. For thousands of years, starlight has aroused curiosity, inspired art, and formed the basis of countless creation stories. Stars have guided exploration, navigation, and astronomy, compelling some of the most significant technological leaps of human innovation.

Night's transformation of the world also serves a crucial biological purpose. Darkness helps regulate sleep patterns in humans and provides migration, feeding, and mating cues for wildlife. Even trees depend on the length of the day to guide the timing of leaf molt. Cycles of light and dark have determined the rhythms of life since life began.

Darkness can evoke fear of the unknown, yet it is also a time for imagination and repose. In turns subtle, intimidating, and spectacular, the night has shaped the patterns of our lives as much as the vibrant world of daylight. For centuries, the world has thrived on this dichotomy. But the night as it used to be is ending—quickly.

In large cities around the world, millions of artificial lights have replaced natural darkness with a pinkish haze that hides all but the brightest stars. Poorly designed fixtures illuminate objects with the intensity of daylight, whitewashing the sides of buildings, streaming through windows and over property lines, and escaping into the atmosphere. Billboard lights and business signs demand attention at hours when there is little likelihood of customers seeing them. Searchlights and light sculptures pierce the darkness, attracting hundreds of birds and night-flying insects. Security lights brashly illuminate lonely stairways, parking lots, and ATM machines, creating shadows that actually reduce visibility.

Where does all this radiance go? Satellite photos taken from NASA space shuttles show the world ablaze with escaping light. This view of earth is simultaneously delightful and daunting. On

The United States at night. COURTESY OF THE DEFENSE METEOROLOGICAL SATELLITES PROGRAM AND NASA

one hand, it is evidence of the presence of life that has the intelligence and determination to alter its surroundings in an exciting, undeniable way. Yet the view also represents great loss: of energy, of exploration of the surrounding universe, and of an essential natural resource.

This loss is caused by light pollution.

Simply put, light pollution is excessive and inappropriate artificial light at night. Light pollution is not *all* outdoor lighting, and important distinctions are made depending on the ecological sensitivity of an area. Artificial lighting is an essential part of modern culture and possibly the most obvious demonstration of technological progress. The problem arises when "progress" crosses the line to become wasted energy, wasted resources, and a negative alteration of the environment.

"Wasted" light shines where it doesn't need to go. Photons that enter windows, invade another's property, or are directed upward to contribute to sky glow are obvious examples. Light that is too bright for a task or shines directly into the eyes is wasted, too. Artificial light, the literal and proverbial "beacon," invites this type of exuberant use. But light pollution is not the inevitable price of progress. It is, in fact, the reverse—a sign of inefficiency and poor use of technology.

Good lighting is important for safety, security, and recreation. But in many cities, outdoor lighting goes well beyond necessity. Ready availability and low cost beg the question "how bright can we get?" instead of "how much do we need?" There is a pervasive attitude that more light is better, even though the opposite is true in many cases.

Light pollution is growing at a rate of six percent each year. Worse, there is no barrier to hinder its spread. The glow over large cities such as Las Vegas sprawls over two hundred miles to intrude on otherwise untouched nightscapes. Areas that offer a pristine night sky are now sites of conservation efforts. In 2007, the European Space Agency's *Rosetta* spacecraft imaged points of light from earth as it flew fifty thousand miles into space to meet a comet. That's the equivalent of light visible at six times greater than the distance to the other side of the earth.

Astronomers were the first to call attention to the purloined night, but they are far from the only group affected. Urban dwellers are subject to loss of sleep from light trespass; small towns and rural areas are constantly getting brighter. Ecological ramifications threaten to disrupt the delicate balance between species, and light shining directly into the sky costs the United States an estimated *2.2 billion dollars* every year in energy consumption.

Another serious issue is the cultural loss caused by the erosion of darkness. For the first time in history, children are growing up without a clear view of the stars. Residents of large cities see fewer than fifty stars in a typical night (in optimal conditions, more than five thousand can be seen with the naked eye). This limited view contributes to a grievous disconnect between science and imagination. Children are inspired by what they experience. Today, many may never actually see a truly starry night sky. The lucky few who do will have to travel ever farther to glimpse what was once ubiquitous. Immediate access to one of the grandest spectacles of the universe—the universe itself—is denied. When the view fades, the interest to preserve it is likely to dim as well. Few who are left will understand the splendor of what was lost.

When Galileo looked up through his telescope—one of the first ever invented—he used a brand new technology to view the cosmos in a way never before experienced by humankind. Since that time, enormous scientific leaps have led to realizations and discoveries that Galileo could hardly fathom. Because of our desire to peer deep into space, we build powerful telescopes that can see distant galaxies. We have even put humans on the moon. It's ironic, then, that the scientific discoveries propelled by that first telescope are pushing technology ahead at an ever-increasing velocity while the cosmos recedes from sight.

But we do have other options.

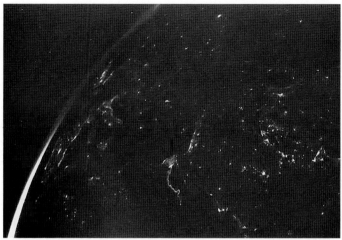

This image of earth, targeted roughly at Greece, was taken with the OSIRIS wide-angle camera by the European Space Agency's Rosetta *spacecraft in 2007. Lights from Europe (far left), on the Arabian Peninsula (right), and winding along the Nile River (center) are seen clearly from a distance of approximately fifty thousand miles.* ESA

As the Milky Way dissolves from the view of two-thirds of the United States and one half of Europe, we face a new dichotomy regarding the night. It is no longer a matter of fear versus inspiration but the struggle to balance the need for darkness with the technology developed to curtail it.

In a 2008 *National Geographic* article, Verlyn Klinkenborg noted, "Of all the pollutions we face, light pollution is perhaps the most easily remedied." Returning the night sky to its natural state is as simple as thoughtful placement of outdoor lighting: putting light where it's needed and darkening areas where it's not. Solutions are available and possible to implement. Moreover, these solutions are guided by principles and technology that save energy, conserve resources, and restore ecosystems. Light pollution is controlled by maximizing efficiency, improving security, and creating a more aesthetically pleasing nocturnal environment. When we eliminate unnecessary light at night in our communities, we also conserve resources, lower costs, and improve the quality of life. If only the other serious environmental issues facing the globe were this easy and rewarding to solve!

The first and best way to do this is simply to look up. If you're one of the lucky few who experience a starry sky, enjoy it—and realize that this vast natural resource needs your protection. If, however, lighting is what you notice, this book gives you the tools to think critically about what you see.

When you look up, you are not alone. Increasing numbers of people are turning their heads toward the night sky. Some are looking for ways to save money or energy. Some realize that an effective ecological conservation agenda requires a natural nighttime environment. Some wonder how the light outside their window is affecting sleep patterns. As the effects of night's absence become known, the cry for its protection becomes louder. The multifaceted issues concerning the loss of the night affect every one of us, and as we begin the second decade of the twenty-first century, public opinion is starting to reflect this fact.

Around the world, public and private groups known collectively as the "dark sky movement" are taking action to return the sky to its natural state. Ordinances against wasteful lighting have increased substantially over the past few years, as has the application of new technology. Awareness is skyrocketing in large cities and small towns, and outreach programs are occurring on a broad scale. Embracing the latest information in optics, vision, and applied science, the dark sky movement uses the technology of today while planning well-designed, resource-savvy cities for tomorrow.

Yet to achieve this requires education and community participation. Individuals and groups must examine *how* we light and *why* we light—and then extend the spirit of invention that created artificial light to implement better ways to use it.

This book offers a comprehensive discussion of light pollution. It examines the serious losses facing society, individuals, and ecosystems as the night fades. It then discusses ways to mitigate them. The last section, on education and conservation, describes paths to progress and how achievements already made toward protecting the natural night can yield significant economic rewards. As knowledge of the issues expands, the reader will understand not only the value of darkness but also the "why" and the "how" to improve sky quality in a community.

One bulb, one fixture, one backyard at a time, the world is beginning to see the darkness. This book offers the tools to become a guardian of the night, reduce energy, protect nocturnal ecosystems, and keep the universe in sight for generations to come.

1

WHAT IS LIGHT POLLUTION?

The universe is filled with light, even in darkness. On a moonless night, earth is illuminated by the light from our galaxy's estimated one hundred thousand million stars. Space dust diffuses the sun's rays to create a distinct nighttime glow known as zodiacal light. Within the atmosphere, particulate matter scatters all available photons to enhance brightness. In such conditions, a fully adapted eye can easily perceive its surroundings.

Yet not all tasks can be performed by starlight, or even moonlight. Humans have sought light to extend activity beyond natural daylight since the first fires were used to heat food. Now most activities in the twenty-first century occur in the presence of artificial illumination. Artificial light allows people to work, to play, to cook, to continue virtually every aspect of diurnal existence at the flick of a switch. And when light is no longer needed, the source of artificial light can be switched off. When the energy that creates the light is discontinued, the light itself disappears.

Light is radiant electromagnetic energy, part of a spectrum of radiant energy traveling in transverse waves at different wavelengths. When this energy travels in a wavelength detectible by the human eye (between approximately 380 and 760 nanometers), it becomes visible to humans as light. Light is not a noxious compound; nor is it a substance. Energy from the sun's light makes life on earth possible. Artificial light, electromagnetic energy generated intentionally through human design, is a practical, even necessary, component of modern existence. How, then, can light be a pollutant?

The Night as an Environment

According to the *McGraw-Hill Dictionary of Scientific and Technical Terms,* "pollution" means "impairment of the purity of the environment." An "environment" is the sum of all external conditions affecting an organism. It is unquestionably established that the state of darkness affects behavior and evolution of all life on the planet. Nocturnal species exist everywhere on the planet, and many primarily diurnal species, from sea turtles to birds, depend on darkness for the performance of biologic activities such as reproduction and migration. Darkness is as significant an environmental quality as light. Therefore, the nocturnal (or nighttime) environment must be established as a unique and essential natural environment. Light can be considered a true pollutant because it affects the natural environment by altering the quantity and quality of total light.

The United Nations Educational, Scientific, and Cultural Organization's Starlight Initiative group developed a working definition of light pollution as "the introduction by humans, directly or indirectly, of artificial light into the environment." While this definition introduces the vital concept of "artificial light" it does not specifically define environment or artificial light. It also presents the idea that all artificial light is a pollutant (and for environmental researchers, this is certainly the case).

Few laypersons, however, would dispute the necessity of creating some artificial light. Artificial light is used in countless beneficial ways. It is therefore necessary to make the distinction between lighting that is carefully designed to minimize stray

Sky glow in New York City. JIM RICHARDSON/NATIONAL GEOGRAPHIC

light while providing adequate illumination for security, safety, and recreation and lighting that degrades or adversely affects the environment.

There are a number of other working definitions of light pollution (as there are for air pollution, water pollution, and other terms). The U.S. National Park Service Night Sky Team is careful to specify light pollution as artificial light as it intrudes into the outdoor environment: principally, the illumination of the night sky caused by artificial light sources, decreasing the visibility of stars and other natural sky phenomena. This definition also includes other incidental or obtrusive aspects of outdoor lighting, such as glare, trespass into areas not needing lighting, use in areas where or at times when lighting is not needed, and the disturbance of the natural nocturnal landscape. Because the team was created to monitor and preserve the quality of the sky over remote public spaces, its definition emphasizes the overall decrease in darkness.

Most definitions of light pollution contain two specific ideas: that only *artificial light* can be a potential pollutant, and that the light must be directed *where it is unwanted, unnecessary, or damaging* in order to be considered problematic. These definitions acknowledge that artificial light designed to minimize impact on the nocturnal environment can be tolerated if it is used appropriately.

The International Dark-Sky Association was created to protect dark skies and control light pollution in all areas, from remote mountaintop observatories to the crowded streets of the world's urban centers. IDA has therefore adopted a definition that identifies the adverse effects of artificial lighting. It defines light pollution as "any adverse effect of artificial light, including sky glow, glare, light trespass, light clutter, decreased visibility at night, and energy waste."

This definition's four lighting components are:

- Urban sky glow: the brightening of the night sky over inhabited areas
- Light trespass: light falling where it is not intended, wanted, or needed
- Glare: excessive brightness that causes visual discomfort and decreases visibility
- Clutter: bright, confusing, and excessive groupings of light sources, commonly found in overlit urban areas

This definition identifies the negative effects of artificial light and describes poor lighting practices for easy identification by people new to the field of light pollution. This definition does not, however, specifically address spectral concerns that are increasingly recognized as affecting human health and biological diversity. As the science behind this field evolves, terms may be adapted or expanded.

Opposite page: Light sculptures such as this one at the Luxor in Las Vegas draw birds, insects, and bats.
JIM RICHARDSON, WWW.JIMRICHARDSONPHOTOGRAPHY.COM

The Cost of Light Pollution

Overall, electric lighting (of all kinds) accounts for 8.3 percent of the primary energy used in the United States.

Primary energy sources, such as coal, oil, gas, and uranium, are converted into useable power: electricity. Electricity use is measured by the kilowatt hour (kWh), or energy consumption (in watts) in one hour's time. One kWh is the electrical energy used by a power flow of 1000 watts for 1 hour or 100 watts for 10 hours.

In 2002, total electricity use in the United States was approximately 3,295 terawatt hours (TWh, equal to 1 billion kilowatt hours). In 2007, that usage had increased to 4,157 TWh.

In 2007, electric lighting used 22 percent of the total electricity generated in the U.S., or approximately 915 TWh.

Of that 22 percent, lighting energy use breaks down as follows: residential (27 percent), commercial (51 percent), industrial (14 percent), and outdoor (8 percent).

Of the total electricity generated in the U.S., outdoor lighting accounts for 1.77 percent (8 percent of 22 percent). This equals 73.2 TWh.

The majority (93 percent) of outdoor lighting is roadway and parking-area lighting. These lamps are the major sources of light pollution. This light comes mainly from the 60 million cobrahead luminaires that constitute 82 percent of all outdoor lighting sockets in use.

Fifty-eight TWh per year is the minimum possible estimate of energy used for outdoor lighting in the U.S. These calculations do not include sports field lighting, floodlit buildings, on-premise signs, decorative lighting, or a neighbor's glaring security light. Many sources of overlighting or light pollution fall into residential, commercial, or industrial categories, which have not distinguished between indoor and outdoor sources. Think also of a skyscraper left blazing after office hours. Energy from these sources is not added to the total figure of 58 TWh.

Total Electrical Energy Use in U.S.
10,840 TWh or 37 quadrillion BTUs (Quads)

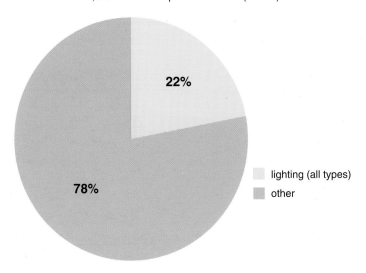

22%

78%

lighting (all types)

other

Breakdown of Lighting Energy Use
by Lighting Type
100% of 22% of total Electrical Energy Use in the U.S.

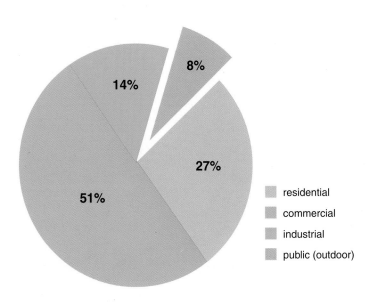

8%

14%

27%

51%

residential

commercial

industrial

public (outdoor)

Percent of Total Electricity Use

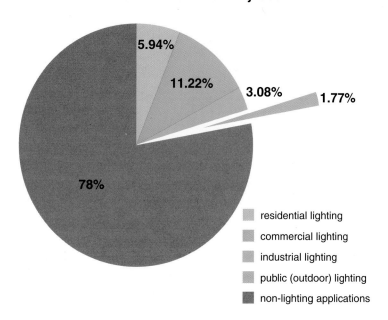

5.94%

11.22%

3.08%

1.77%

78%

residential lighting
commercial lighting
industrial lighting
public (outdoor) lighting
non-lighting applications

Breakdown of Public (Outdoor) Lighting

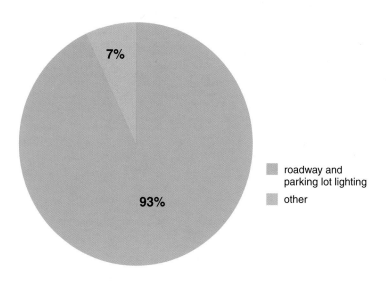

7%

93%

roadway and
parking lot lighting
other

Photometric analyses of luminaires have determined that "wasted light"—light that is not reflected but sent directly into the sky—from outdoor lighting remains constant at, on average, 30 percent of the total light emitted by public lighting. This amounts to 22 TWh (2.2 billion kWh) each year of wasted energy.

At 10 cents per kWh (approximately the national average), the cost of that wasted energy is *2.2 billion dollars.*

The average American household uses 1,946 kWh a year in lighting energy. The uplight from outdoor public lighting fixtures—light that essentially shines up into the sky—could light more than 11 million homes every year.

One tree on the equator absorbs 21 kilograms of carbon dioxide annually. It would take 702 million trees to absorb the 14.7 million tons of carbon dioxide produced by wasted light.

An Important Note

This analysis acknowledges that energy use values from 2007 are likely low, but the percentage breakdowns (such as the 8 percent value for outdoor lighting energy use as a percentage of all lighting) have remained reasonably constant.

Much of the wasted light in this estimate comes from public roadway lighting. As the cost of energy increases, local and state governments are considering ways to reduce energy costs through lighting switch-offs or retrofits. States such as Connecticut and New Hampshire have reaped savings by seeking a special part-night utility rate from electric companies for lights that dim or shut off at midnight. Though many options for saving money through reduced lighting exist, some choices are far more effective and environmentally sound than others. Many factors to be considered are discussed in this book.

While the increased use of LED technology has the potential to significantly alter current figures, at this point, no widespread data on LED lighting usage have been collected. This rapidly developing technology offers many exciting improvements in efficiency, as do mechanical advancements such as better refractors to precisely direct light. LED technology, or any other technology, must not be considered a solution to light pollution, however.

Efficiency improvements alone are not sufficient to make a large impact on the quality of the nocturnal environment. Such improvements do not address the important question of spectrum distribution or light quality; nor do they address the human propensity to use more of a resource when the cost decreases.

The truth is that lighting technology by itself will not decrease light pollution. In fact, as efficiency increases, lighting designers are noticing an inclination to add more light to a site in response to the reduced energy costs. Using light efficiently by considering lamp placement for best visibility, lighting to minimum specifications, controlling the spectral quality of light, and illuminating only what is needed is necessary to restore the nocturnal environment and decrease energy usage, cost, and carbon footprint.

2

DISABILITY GLARE

One of the basic problems caused by light pollution is the creation of glare. Disability glare is unwanted and poorly directed light that blinds, causes poor vision by decreasing contrast, and creates unsafe conditions, especially at night. It is a common feature of poorly designed lighting systems, especially along roads and highways. And it is a particular problem for older people. In fact, many elderly drivers have a difficult time driving at night, but are unaware of the underlying cause of their poor night vision: badly engineered lights.

There are natural causes of disability glare, such as sunshine on a dirty windshield at dusk. We have all experienced such glare and probably attempt to minimize it with sunglasses or by wiping off the windshield. Unfortunately, glare at night while driving is not so easily remedied. Its cause is usually overly bright and unshielded or poorly directed light that enters the eye and then scatters off eye structures, resulting in diminished contrast and impeded vision. These effects become dramatically worse as the human eye ages.

The human eye is a marvel of evolutionary engineering, and it functions in much the same way as a well-made camera. Light enters through the eye's clear cornea, where initial crude focus occurs. Corneal imperfections can scatter light, however, preventing a clear image and lowering contrast. Light then moves through the anterior chamber and passes through the iris, an aperture that has the ability to contract to one millimeter in diameter and open to about seven millimeters.

This dynamic structure reacts to both the ambient light level and any bright source of light. In dimly lit surroundings, the iris relaxes, allowing up to thirty times more light to enter than in brightly lit surroundings; in bright light, it closes down to protect itself from bright-light effects. At night, when the pupil should be large to allow the maximum amount of available light to enter so night vision can work properly, light from unshielded, bright light sources forces the iris to contract. The result is that the eye can be effectively up to ninety-six percent closed just when seeing clearly is especially important.

After the iris, light passes through the eye's lens. This remarkable organ is under the control of tiny muscles that alter its shape and allow for extremely fine focus both near and far. As the eye ages, however, this lens grows layers (much like an onion) of clear cells. Generally, by age fifty, the lens becomes so thick that it loses the ability to adapt to both near and far vision. As well, the central cells are deprived of nutrients and blood vessels and eventually die, yellowing and crystallizing. This can create opacities and cataracts, which scatter and absorb light. With the lens thus affected, any bright beam of light, even if it's off-axis from what the eye is focused on, will be scattered, creating a whiteout and loss of contrast. It's like being in a blinding snowstorm.

After the lens, light passes through the vitreous humor, a jellylike substance that fills the main eyeball. Here, dead and sloughed-off cells from the retina accumulate (these are the so-called "floaters" that are a natural part of the aging eye). These cells can further scatter light and diminish contrast. Light eventually strikes the retina, which lines the back of the eye. Here, the retina is composed of millions of tiny cells that convert a photon of light into an electrical signal that is interpreted by the human brain as an image.

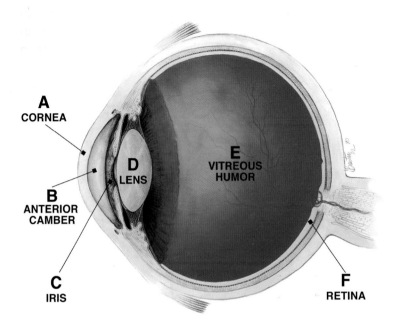

Diagram of the human eye. NATIONAL EYE INSTITUTE, NATIONAL INSTITUTES OF HEALTH (NEI/NIH)

There are four kinds of receptors in the retina. Three types of cone cells react to the colors red, green, and blue. They are responsible for all color vision and are fully activated in bright sunlight. Cone vision is known as *photopic vision*. A second type of receptor, a rod, is designed primarily to react to low-light conditions, such as moonlight. Rods are used in *scotopic vision*. There are many more rods than cones, and they are so sensitive that their function is impaired by the bleaching caused by a bright light source, such as poorly designed streetlights or spotlights that create high contrast between light and dark.

While impaired rods can recover relatively quickly, glaring streetlights are often in a repetitive pattern, with a bright light source every hundred feet or so. In these conditions, the rods never recover fully: they simply don't have enough time to react before they're hit with more glare. Older drivers are particularly affected by poorly designed streetlights, because their rods take longer to recover, but glare affects everyone to varying degrees.

Preventing Glare

With a proper understanding of how the human eye focuses, images, and reacts to various lighting conditions, and of the various pathological conditions that develop with aging, certain principles of good lighting engineering become apparent.

Direct high-intensity beams should never be shined into the eyes. All street lighting should be shielded so no direct light from the bulb acts as a bright "hot spot." Decreasing direct visual contact with a light source minimizes glare, improves visual contrast, and helps the eye's ability to adjust to low lighting. Following this simple principle would allow for pupil dilation, maximizing vision in dim areas. And it would prevent bright light from entering enter the eye and scattering, which diminishes contrast and impairs vision further.

Indirect lighting should be used as much as is practical. We see objects not by a nearby light fixture but by the light that reflects off them. This principle is key to understanding vision. When a camera captures an object lit by a bright, direct light, the image appears washed out, and details are lost because of high contrast. Professional photographers know to use ambient light or indirect lighting to render subjects accurately and accentuate detail. Visual perception in the human eye works much the same way. Details can be more easily perceived when a subject is not overlit.

Less lighting in dark environments can improve visual capacity. If they are not subject to constantly changing light levels, rods in the eye can function to their fullest potential. Even if it doesn't mimic daylight and offer a full range of color, a constant low level of lighting gives the eye time to adjust and make use of *all* available light.

3

EFFECTS ON HUMAN HEALTH

Humans, like many other organisms, have evolved in the presence of the daily light-dark cycle caused by the earth's rotation. This twenty-four-hour light-dark signal has provided a powerful evolutionary pressure for adaptation to particular temporal niches: to being day-active (diurnal), night-active (nocturnal), dawn- and dusk-active (crepuscular), and many variations. Many specialized adaptations have evolved to optimize physiology and behavior in relation to the time of day, solar orientation, season, and the amount of light or dark in the environment.

It is only relatively recently that humans developed the ability to generate light—about 250,000 years ago, with the use of fire. Candles were developed about 5,000 years ago, and gas street-lighting was possible beginning in the mid-1700s. In the last 120 years, however, our ability to alter our environment has expanded dramatically with the creation of the electric light. This light affects all organisms exposed to it, not just humans, and the consequences of such a profound alteration in one of the most powerful environmental signals are not yet fully known. We are only beginning to understand the impact of artificial light on human health. Research over the past century, however, has shown that light exerts powerful effects on human physiology, endocrinology, and behavior. Because humans evolved in a distinct light-dark cycle, it is likely that unnatural exposure to artificial light at night poses hazards to human health.

Photoreception in Humans

Before you can understand the potential risks that exposure to light at night might pose, it is necessary to understand some of the basic mechanisms through which light affects human physiology. Light has myriad effects in humans because it is mediated by photoreceptors that are tuned to detect and transduce specific wavelengths of light to induce specific responses. Photoreceptors contain the molecules, or photopigments, that detect light. These have a specific light color detection "fingerprint" closely matched by the spectral sensitivity of the biological responses mediated by the photoreceptor. Photoreceptors can be present in many types of tissues. In humans, for example, photoreceptors in the skin detect the ultraviolet light necessary for the synthesis of vitamin D. Most of the effects of light in humans, however, are mediated through photoreceptors in the eyes. Vision, considered the predominant function of the eye, is mediated by rods and short-, medium-, and long-wavelength visible-light cone-photoreceptors. These detect dim light and colored light, respectively.

In 2000, a different photoreceptor system was discovered in the mammalian eye, one that is anatomically and functionally different from the rod-and-cone photoreceptors used for vision. This newly discovered system detects light for a range of "nonvisual" responses. The system's photoreceptor cells are located in a different layer of the eye than are rods and cones—in the ganglion cell layer. And they contain a newly discovered photopigment called melanopsin. Unlike normal ganglion cells, which are stimulated by light indirectly by signals from rods and cones, melanopsin-containing ganglion cells are *directly* photosensitive.

The cells are small in number (about three percent of the total number of ganglion cells) but spread across the retina in a network to cover the entire visual field. Such a distribution is ideal for detecting general light irradiance and changes in light according to the time of day and season. In fact, such general light detection is more fundamentally functional than detailed vision; even single-celled organisms with it are capable of detecting the twenty-four-hour light-dark environment.

So what does this novel photoreceptor system do? Much like the ear has dual functions—both hearing and balance—the human eye has dual roles: seeing and detecting light for a range of behavioral and physiological responses. Light exposure in the eye induces multiple neuroendocrine, neurobehavioral, and physiological responses: suppression of pineal melatonin production, constriction of the pupils, heart rate and temperature regulation, enhancement of alertness and performance, changes in brain activity patterns, phase-shifts of the internal circadian pacemaker, and even stimulation of circadian-clock gene expression.

For want of a better expression, these wide-ranging effects of light are collectively called *nonvisual responses.* They are sometimes grouped under the term "circadian photoreception," as much of the behavioral and neuroanatomical work that first identified them was focused on studies of the ability of light to shift the timing of the circadian pacemaker. While our basic understanding of these responses has been derived largely from experiments in rodents, parallel studies in humans have characterized similar properties of the system.

Physical Effects of Light

How do the nonvisual effects of light affect the physiologies of humans and other animals? First, many aspects of human physiology, metabolism, and behavior are dominated by twenty-four-hour rhythms that have a major impact on our health and well-being. For example, sleep-wake cycles, alertness and performance patterns, core body temperature rhythms, and the production of hormones such as melatonin and cortisol are all regulated by a near-twenty-four-hour oscillator in the brain. Oscillator cells generate rhythms with a period close to, but not exactly, twenty-four hours (circadian means "about a day"). In order for the circadian pacemaker to ensure that physiology and behavior are appropriately timed to anticipate events in the outside world (the early bird must be awake before the worm in order to catch it), environmental time cues must be able to reset this internal clock.

The major environmental time cue that resets these rhythms in mammals is the twenty-four-hour light-dark cycle. Findings indicate that inappropriate light exposure can cause circadian rhythms to become desynchronized—both from the environment and from each other—with potential adverse effects on human physiology and metabolism.

Light also affects the major biochemical signal for darkness, which is provided by the pineal melatonin rhythm. Under normal light-dark conditions, melatonin is produced by the pineal gland only during the night. It provides an internal representation of the environmental night-length. The synthesis and timing of melatonin production requires a signal from the brain's oscillator that projects to the pineal gland. Disruption in the twenty-four-hour cycle disrupts the oscillator's function.

Light exposure during the night inhibits melatonin production. Melatonin suppression occurs immediately upon light exposure and stops when the light is turned off. Under a natural light-dark environment, melatonin production would not occur when it was light and so would not be suppressed. But if prolonged exposure to light during the melatonin-secretion phase at night is a daily occurrence, regular, chronic melatonin suppression is the result.

Finally, light exposure at night is capable of directly elevating heart rate and core body temperature and affecting cortisol production. While these acute effects of light are relatively small and transient, the long-term consequences are unknown.

Circadian Rhythm Sleep Disorders

The best-established disorders caused by inappropriate exposure to light are circadian rhythm sleep disorders. These disorders are commonly caused by exposure to light at inappropriate times— during the night. The most common chronic circadian rhythm sleep disorder is shift-work disorder, experienced by many of the estimated fifteen million shift-workers in this country. Shift-workers experience more sleep problems, fatigue, forgetfulness, performance problems, gastrointestinal problems, and greater incidence of accidents and injuries than do regular-workday workers. They also have an increased long-term risk of cardio-vascular disease, type-2 diabetes, and some types of cancer.

The underlying source of these problems is desynchroniza-tion between the shift-work schedule and the light-dark cycle. Circadian desynchronization is caused by a failure of the inter-nal circadian pacemaker to remain determined by the environ-mental light-dark cycle. Shift-workers' schedules (and therefore light-dark cycles) often change more rapidly than the circadian system can, resulting in wake and sleep occurring at an inappro-priate circadian phase. Consequently, daytime sleep duration and quality is reduced, and workplace alertness and perform-ance are impaired, increasing the risk of accidents and injuries. (Jet lag is essentially the same problem but is usually an acute, rather than chronic, desynchronization.)

While the precise mechanisms of these long-term health risks are not known, there are an increasing number of studies detailing the detrimental effects of light, sleep restriction, and circadian desynchronization that might underlie them.

Shift-workers often eat at an inappropriate circadian phase, impairing metabolism and creating higher-than-normal levels of post-meal insulin, glucose, and fats. Recent studies have shown that cells in individual organs such as the liver, heart, lungs, and kidneys are also capable of generating internal circa-dian rhythms. It is likely that synchronization of these internal rhythms is at least as important, if not more so, as synchronization with external cycles in the maintenance and complementary timing of metabolic processes. Within the circadian system, metabolic processes are not as efficient at some times of day—the chances of getting indigestion after eating a pizza at 3 A.M. are greater than if it were eaten at 3 P.M. The chronic elevation of insulin, glucose, and fats observed in shift-workers who eat at the wrong time of day are symptomatic of insulin resistance and metabolic syndrome and are risk factors for type-2 diabetes and cardiovascular disease.

Reduced sleep duration may itself play a role in altering metabolism. Studies have shown that sleep restriction can cause an imbalance of two important hormones, leptin and ghrelin. Leptin is a satiety signal; increased production of it decreases appetite and food intake. Ghrelin stimulates food intake. Experimental sleep restriction causes leptin to decrease and ghrelin to increase, increasing appetite and food intake. Epidemiological studies show that short sleepers have reduced leptin, elevated ghrelin, and increased body-mass index compared to longer sleepers. Similar links have been shown in studies of young children who sleep fewer than twelve hours. They have a higher risk of being overweight. Collectively, these data suggest that eating meals during the biological night and not sleeping enough elevate risk of metabolic and cardiovascular disorders.

Cancer Risk

The relationship between shift-work and cancer risk has received much attention of late. In 2007, the World Health Organization stated that "shift-work that involves circadian disruption is probably carcinogenic to humans." Several high-quality epidemiological studies (although not all the studies) have shown that female shift-workers have about a fifty percent increased risk of breast cancer, and some (but again not all) support similar findings for colorectal cancer risk in women and prostate cancer risk in men.

(Female flight attendants also have an increased risk of breast cancer, illustrating the potential similarities between shift-work and jet lag.) While the studies cannot address the mechanism causing the disease, several hypotheses have been put forward.

Given that shift-workers by definition are often awake at night, one hypothesis proposes that light exposure at night is a potential mechanism for the increased cancer risk. This hypothesis (first proposed by cancer epidemiologist Richard G. Stevens) is based on the finding that cancer rates increase as nations become more industrialized—increased artificial lighting is a common consequence of industrialization. This idea is also based on some of the physiological effects of light in relation to cancer risk in animal studies and in vitro. One of the effects of light exposure at night is to suppress production of the hormone melatonin. In animal models, suppression of melatonin by exposure to constant light, or by the removal of the pineal gland, will increase the development of mammary tumors in rodents. Melatonin administration will inhibit the proliferation of human breast cancer cells in culture; it will also inhibit liver cancer growth in rats.

A recent study of human breast-tumor growth after implantation in an immuno-suppressed rat showed that infusion of human blood high in melatonin was able to slow tumor growth; infusion of blood low in melatonin (drawn after exposure to light at night) was unable to prevent tumor development. Melatonin is a potent free-radical scavenger that might also play a role in preventing cancer cell damage and proliferation.

While these data suggest a strong association between light exposure, melatonin, and cancer, there is as yet no direct evidence in humans proving that alteration of melatonin levels alters cancer risk, or that taking synthetic melatonin has any effect on cancer risk or proliferation. Work is continuing to explore these relationships.

Another hypothesis about what underlies the increased cancer risk associated with shift-work is disruption of circadian rhythms by exposure to unusual light-dark cycles. As discussed, connection of the circadian rhythms with the outside world and with each other is required for normal biochemical and physiological function. The "circadian clock genes" that generate spontaneous circadian rhythms in the brain are also present in many peripheral organs and cells. Animal and in vitro studies have shown that, while light is the major environmental time cue for resetting the clock, peripheral organs such as the liver, lungs, and heart may be more sensitive to nonlight time cues, such as meals. It makes intuitive sense that the circadian clocks in the esophagus, stomach, intestine, liver, kidneys, and bladder must be synchronized well with each other to optimize their function. Shift-work (and jet lag) involve the disruption of both light-based and nonlight time cues and will alter the relationship between these internal rhythms. Indeed, studies comparing rodents living in a regular light-dark cycle with animals placed on a shift-work or jet-lag-type light-dark schedule show higher rates of cancer initiation and progression in the irregular group.

Continuing Research

While inappropriate light exposure and circadian disruption due to shift-work are well defined, the effects of inappropriate light exposure while living on more regular schedules are only just starting to become known. Our modern sleep-wake schedules are far removed from the natural light-dark cycle—"midnight" is certainly not the middle of the night anymore—and the full consequences of our ability to light the night and be awake and active later and longer are not yet apparent.

Given that dim light is capable of stimulating effects on human physiology, we cannot consider dim light exposure at night an inert stimulus, whether we're awake or asleep. We must keep this in mind when reviewing the appropriateness of light environments at night. Studies are underway to measure the actual light levels that people are exposed to while indoors.

In urban environments, these light levels are likely to be significant, and even higher when individuals live close to intrusive street lighting. Unnecessary horizontal and vertical street lighting permeates living spaces, particularly bedrooms. This light intrusion, even if it's dim light, is likely to have a measurable effect on sleep and melatonin suppression and other physiological and metabolic processes. Even if these effects are relatively small from night to night, continuous, chronic disruption may have longer-term health risks. Measures to reduce light pollution are very likely to have a beneficial effect on human health.

Local, state, and federal governments must learn about these potential risks and take advantage of the potential health and energy benefits of reducing, or eliminating where possible, light at night. Some have done much: Based on the WHO categorization of shift-work as a probable carcinogen, the Danish government has paid compensation to female shift-workers and flight attendants who developed breast cancer.

More education on this issue is needed. Everyone should be made aware of the potential health risks posed by artificial night lighting. It is ironic, for example, that the Estee Lauder Companies' Global Landmark Illumination Initiative—a program that focuses attention on breast health and spreads the message that early breast-cancer detection can save lives—does so by bathing famous landmarks across the world in pink light at night.

4

EFFECTS ON WILDLIFE

A growing body of data suggests that artificial lighting at night has negative and deadly effects on a wide range of creatures, including amphibians, birds, mammals, insects, and even plants. All animals, not just humans, depend on a regular interval of daylight and darkness for proper functioning of behavioral, reproductive, and immune systems. For thousands of species, the dark night of the evolutionary past is an integral part of their existence. The disappearance of true darkness portends grave consequences for these creatures.

Artificial night lighting affects not only nocturnal creatures but also diurnal species, active during the day, and crepuscular species, awake at dawn and dusk. The introduction of artificial light into an ecosystem can harm species directly in a variety of ways. It triggers unnatural periods of attraction or repulsion that lead to disruptions in reproductive cycles. It fixates and disorients animals. It interferes with feeding and sustenance. Studies have shown that artificial light at night affects the movements of migratory birds and hatchling turtles, disrupts mating and reproductive behavior in fireflies and frogs, and interferes with communication in numerous species, from glowworms to coyotes.

It also harms species indirectly. The degradation of habitat, creation of artificial and dangerous habitat, and energy waste that may lead to climate change can all be linked to excessive artificial night lighting. Research biologists warn that the negative synergy of such combinations can result in a cascade effect, with disastrous results for entire ecosystems.

The delicate balance of interspecies interaction can be upset when outdoor lighting artificially extends the length of daylight. Photoperiod—the recurring cycle of light and dark during a twenty-four-hour span—has for millennia been a dependably consistent factor in the natural environment. Climate characteristics vary from one year to the next: It is not uncommon to experience cool summers and lingering autumns. Therefore, many species of plants and animals rely on the length of the day to indicate the proper season for mating, molting, and other activities. Photoperiodic sensitivity is so acute that many species can detect discrepancies in the length of the day as short as one minute.

The full extent of the consequences of photoperiodic alteration by artificial light is not yet known. Reproduction cycles are disrupted when artificial light at night interferes with plants' and

animals' light-detection systems. Trees bud prematurely or fail to lose their leaves when given false cues from outdoor night lighting, and some flowers stop blooming. Squirrels, robins, and other species may mate out of season. These individual changes may seem inconsequential, but each change in timing desynchronizes the finer points of a complex and interdependent ecosystem. For instance, migration paths of birds and bats are timed to coincide with the blossoming of edible trees and flowers along migration routes. This beneficial symbiosis feeds animals and pollinates plants. A tree budding prematurely is not a viable food source and is less likely to become pollinated. Thus, minute changes in plant and animal reproductive activity cause problems not only for animals that are directly affected but also for other organisms they interact with.

Artificial light affects reproduction in other ways, too. Amphibians seem especially prone to reproduction disruption because of artificial night lighting because they must remain close to a water source and thus are less able to compensate for changes in the environment by relocating. Many pond-breeding salamanders show a strong fidelity to their home ponds, and studies have shown that artificial illumination can disrupt salamanders' ability to return to home ponds to breed. Like other amphibians, salamanders are currently suffering population declines around the world.

As well, firefly populations are decreasing noticeably, and observations suggest that artificial light at night interferes with numerous firefly behavior mechanisms. Fireflies respond to light cues in flight altitude and flash pattern. Flight trajectory and their own pattern of bioluminescence have been linked to feeding and reproduction. An artificially brightened environment may cause some behaviors to be artificially triggered and others to stop, possibly decreasing the total time an adult spends attracting a mate. Artificial light also makes the insects more visible to predators and decreases the effectiveness of flash patterns by making mating behavior less visible.

Opposite page: Light from development on beaches confuses nesting sea turtles, which have returned to these shores for generations. JIM RICHARDSON

Artificial light at night can contribute to animal starvation by interfering with predator-prey relationships. For instance, many moths and other night-flying insects are automatically attracted to lights. This involuntary movement leads to their easy capture by predators. Moths' incessant gravitation toward artificial points of light disrupts the normal nocturnal patterns of predator species by creating an artificial concentration of food. For some predators, such as bats or birds not repelled by light, this disruption can create a dramatic change in the quantity and location of their food, which can lead to imbalances in predator-to-prey ratio—with negative consequences for both species.

Since the eyes of nocturnal animals are specially evolved for foraging in low-light conditions, small changes in illumination can compromise strategies and profoundly alter their relationship with prey species. For example, some species of nocturnal fish respond to extremely low illumination levels. Artificial light can induce predator species to forage. This altered behavior may create an advantage for some, but the advantage of one species can be detrimental to other species that don't or can't alter their behavior.

Ongoing experiments show how artificial light can affect bats. Foraging brown long-eared bats rely more on visual than

Glaring lights on oil platforms attract many types of marine life.
WWW.OFFSHOREPICTURES.COM

acoustic cues, suggesting that they prefer to locate food by sight. For long-eared bats and other species repelled by light, such as horseshoe and mouse-eared bats, food becomes scarcer and difficult to procure when insects swarm around lights, leaving fewer to be caught elsewhere. Current studies of the United Kingdom's endangered lesser horseshoe bat show that the species will take a longer route to feeding grounds in order to avoid lighted areas. That longer commute means that the bat expends more energy, is exposed to predators longer, and has a shorter time to feed. Moths are a favorite food of lesser horseshoe bats, but when moths are drawn to the very light sources these bats avoid, they cease to become a viable prey for the hungry mammals. The decreasing amount of available food, due to a combination of habitat loss and life-cycle disruption, is contributing to an overall decline in many bat populations.

Like moths, some species of fish, normally exposed only to natural light sources such as phosphorescence, can be dangerously drawn to light sources. Large night-fishing enterprises use light to attract fish and squid species. Other ocean dwellers can be temporarily blinded and left vulnerable by this artificial light, which also inhibits normal antipredation behavior such as schooling and can affect migratory patterns in fish species such as salmon, specifically sockeye fry.

Special Problems for Birds

Artificial light at night presents special difficulties to migrating birds. Though most migratory birds are diurnal, hundreds of species migrate at night. In the northern hemisphere, millions of individuals make seasonal journeys, flying north from mid-March to early June and south from early August to early November. They are guided in part by the moon and constellations. If the skies are clear and their path is moonlit, migrating birds fly at high altitudes. Fog, rain, or low clouds will bring them much closer to the ground, however, as will preparations to land to rest and feed. It is then that birds are particularly vulnerable to dangers posed by light at night.

When birds fly at low altitudes, especially during fog, they will often fly directly into the lit windows of skyscrapers, which kills some outright and stuns many others. Some of the stunned birds revive in a couple of hours but then may become trapped in a maze of bright, reflective buildings. Bright light sources are powerfully hypnotic to birds. Disoriented individuals will fly around a single light source, circling until they drop, exhausted. Stunned or exhausted birds can fall victim to scavengers such as gulls, crows, raccoons, and cats.

Native songbirds—not the alien species such as starlings that are overtaking crucial habitat—are the most common victims of collisions. These birds are critical to a healthy environment. They eat billions of insects every year, pollinate plants, and disperse seeds. They are also valuable to our economy—bird-watching has become one of the top leisure activities in North America. And they are arguably priceless simply for what they are. Their beauty, songs, and intimate connection with wilderness are irreplaceable and enrich our lives immeasurably.

Centuries of anecdotal evidence have proven how lethal the combination of night lighting and bad weather can be to birds—historically (and to this day) at lighthouses and, more recently, at emission stacks, ceilometers, tall monuments, and other towering structures. On nights when huge flocks encounter foggy, cloudy, or rainy weather in the vicinity of such structures, there can be a spectacular loss of life. Over two nights in September 1968, for example, more than five thousand warblers and other birds were killed at a television tower in Nashville. In October 1950, at a ceilometer on Long Island, two thousand blackpoll warblers died. An estimated fifty thousand birds were killed over two nights in October 1954 at Robins Air Force Base in Georgia. During fall and spring migration seasons from 1960 to 1969, almost seven thousand dead and dazed birds were found beneath the lighthouse at Long Point, Ontario.

Hundreds of avian victims of building collision in and around Toronto, Ontario, were collected by volunteers in 2009. KENNETH HERDY/FLAP

Devastating Effects on Turtles

Like migratory birds, sea turtle hatchlings are devastated by artificial light at night. Immediately upon a nocturnal entry into the world, a sea turtle hatchling must accurately locate an ocean it has never seen and begin a swim for its life. During this emergence event, a little turtle is among dozens of its siblings erupting from a sandy depression above their buried empty eggshells. As they scramble to the vantage point atop the weathered mound left by their mother, each takes in its first scene—a starry night sky over two opposite horizons. Choosing between the two horizons is a decision that will either lead a hatchling safely from nest to sea or condemn the little turtle to death within a tangled dune.

This biological drama has taken place for millions of years and on widely varying beaches all over the world. How is it done? How do hatchlings make the correct orientation choice? At a basic level, we understand how hatchlings point themselves seaward: by orienting toward the center of a bright, broad, unobstructed horizon. These characteristics typically match the open view of the night sky over a glittering sea. On beaches lit by only the twinkle of stars or a glowing moon, hatchlings crawl on precise bearings toward the seaward brightness. But on beaches lit by the glare of electric lights, hatchlings often attempt to reach the artificial brightness that overpowers the subtle cues from natural nocturnal light. So deceived, a hatchling is unlikely to ever reach the sea.

When hatchlings emerge on a beach with no visible artificial light, it takes only a few minutes for them to crawl to the safety of the ocean. On naturally lighted beaches, hatchling tracks from a nest fan out about 45 degrees, all showing more or less a straight path toward the water. Tracks of hatchlings exposed to artificial lighting often span from 90 to 360 degrees. Many hatchlings turn around in circles or walk parallel to shore. Hatchlings that can sense an artificially lit horizon have been known to cross roads and encounter myriad dangers—including dehydration—that severely decrease their likelihood of survival.

Proper light management gives hatchling turtles a chance to orient directly toward the water.
BLAIR WITHERINGTON

Female sea turtles seeking nesting sites are also affected by light pollution. In most species, this important reproductive activity takes place only at night, with the darkest areas of beach most favored for nesting. Adult turtles can also have their orientation disrupted. These adult turtles, weighing up to 270 pounds, are struck by cars on lit roadways or drawn away from water sources.

In Florida, light pollution on sea turtle nesting beaches is a serious conservation problem. Although the toll from uncontrolled lighting on sea turtles is difficult to estimate, it is likely that this problem results in the deaths of hundreds of thousands of hatchlings each year in the state. Florida hosts one of the two largest loggerhead populations in the world and the second largest green turtle population in the wider Caribbean.

Though sea turtle reproductive research has been conducted for over a century, only about fifty years ago did the connection between light cues and hatchling orientation become clear. Today, the disruptive effect of artificial lighting on sea turtles has been extensively demonstrated. In addition, there has been considerable research on how and why sea turtle hatchlings orient using light, work that has been instrumental in guiding efforts to manage light pollution near nesting beaches.

For the most part, ongoing studies reveal that most turtles orient toward the brightest direction measured horizontally wide and vertically flat, showing that light closest to the horizon plays the largest role in determining hatchling orientation. The draw of wide horizontal light also shows that light reaching the hatchling (irradiance) is more important than light emanating from a source (radiance). Thus, even when a celestial light source such as the moon is opposite the ocean, its less directional light field will not affect a hatchling's ability to orient itself the way a highly directed light field from artificial sources would.

Hatchlings are also sensitive to light spectra. Some wavelengths of light are very attractive to hatchlings, while other wavelengths seem to have little to no effect on orientation.

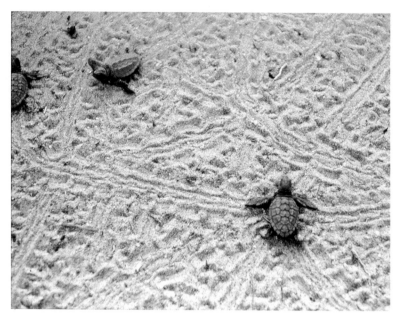

Disoriented by light, newly hatched turtles are uncertain as to which direction the ocean lies in. FLORIDA FISH AND WILDLIFE COMMISSION

Orientation trials featuring hatchlings that were shown light through narrow-band filters indicate that shorter wavelengths (bluer light) are much more attractive than longer wavelengths (redder light).

This information is helpful in determining what types of light will have the least impact on hatchlings. Despite the complexity of understanding how artificial contributions to the light fields affect sea turtles, one simple rule has proven useful in identifying light pollution problems on beaches—*any artificial source producing light that is visible from the beach is likely to cause problems for nesting sea turtles and their hatchlings.* The rule is helpful because nearly all artificial sources visible to a sea turtle would also be visible to a human.

Of course, turning off all beachside lighting in Florida, where most of the state's population is concentrated on the coast, is not an option. Solutions have been found, however, by using best available technology (BAT) and best management practices (BMP) borrowed from air- and water-pollution management models.

Some of these lighting technologies and practices are simple indeed—installing shields, directing light downwards, lowering lights to diminish their footprint, and using low-tech methods to hide light sources from the beach. Other technological solutions come from choices determining spectral emission. Yes, some light sources are more harmful than others, although no visible light is completely harmless. Exclusively long-wavelength lamps such as low-pressure sodium vapor and red and amber LEDS are the least harmful and the best choices when used in fixtures that are mounted low, direct light downwards, and control light well.

These principles can be applied to combat ecological impacts of artificial lighting in many areas and have some application in the protection of other species as well. Another important aspect of wildlife conservation is to keep habitat as natural as possible. In addition to reducing light, preserving native flora and natural landscape features will provide more stability for all wildlife affected by human development.

Case Study: Florida

With its burgeoning coastal population and importance to nesting sea turtles, Florida has been presented with a compound and growing set of light-management challenges. Meeting these challenges has been an ongoing effort for numerous coastal communities and two state agencies, the Fish and Wildlife Conservation Commission (FWC) and Department of Environmental Protection (DEP). Local ordinances and agency rules have been instrumental in this effort, with an initial promulgation in 1985 when Brevard County passed the first light-management ordinance for sea turtles in the nation. Following suit, twenty coastal communities and fifty-seven coastal municipalities passed similar ordinances, which now cover nesting beaches for the majority of sea turtles in the state.

FWC assisted in these local efforts by publishing a model lighting ordinance to guide effective legislation. The agency's role has also included advising DEP on permitting new construction, producing and distributing guidance documents, convening workshops, and monitoring lighting effects on sea turtle beaches.

Educational efforts in Florida have also become widespread. Much of this education is credited to small nonprofit groups and at least one power company—Florida Power and Light. In their role, lighting distributors conducting business in Florida now seek a "Wildlife Friendly" certification for their lamps and luminaries, which indicates how well a light source fits "best available technology" for sea turtles and other light-sensitive animals.

STORIES OF DARKNESS

Paul Bogard

In the late 1920s, as he witnessed the encroachment of artificial lighting into the Cape Cod area he loved, Henry Beston wrote, "With lights and ever more lights, we chase the holiness and beauty of night back to the forests and the sea." Were Beston with us today, he would no doubt understand and lament the costs of light pollution—the waste of energy and money, the threats to human health and safety, the impact on ecosystems worldwide—but my sense is that he would mourn the effects on "holiness and beauty" the most. As a writer and storyteller, Beston knew, as generations of storytellers before him knew, the vital importance of night to the human imagination, to the human soul. He knew that the price we pay for "lights and ever more lights" can't always be quantified by dollar signs or scientific studies.

The experience of real darkness has shaped human culture. From our first storytellers to our most recent—around campfires, in the theater, on canvas, on the page; in mythology, religion, poetry, painting, literature—the influence and inspiration of night have been steady and consistent. Simply put, the loss of darkness to light pollution threatens a quiet, irreplaceable part of what it means to be human.

In the summer, I do my best stargazing and moon-watching from the end of a dock on a lake in northern Minnesota. My parents built our cabin here the year I was born, and I have spent parts of every summer for forty years here with the night sky. While the usual threats are clearly gnawing at the horizon, the sky above the lake remains quite beautiful, and real darkness has been part of my experience here all my life.

Often when I'm on the dock or in our ancient canoe after midnight, stars spread above, sometimes with a waxing or waning gibbous moon lifting slowly from the horizon, I think of those who have witnessed a similar scene before me. Sometimes I imagine my parents and grandparents here before I was born. Sometimes I imagine an Ojibwa man my age here a hundred and fifty years ago. In fact, at night on the lake, it's easy to step back in time to imagine any human gazing at the night sky. I know that in what I'm doing tonight—gazing with wonder at the stars and the moon, feeling the impulse to scribble on the notepad in my lap—I join in one of the oldest of human experiences: being inspired by the night's darkness.

This inspiration comes in many forms. We usually think of it in terms of Beston's "beauty," of standing beneath a starry night and feeling a sense of wonder and awe and excitement that might lead an artist to react through creating—think of Vincent Van Gogh being inspired to paint *Starry Night* (one of many night paintings) or, in literature, entire chapters, such as Beston's "Night on the Great Beach" from his book *The Outermost House.* Or scenes like those of Mark Twain taking Huck and Jim down the Mississippi at night. Or single lines like F. Scott Fitzgerald's "silver pepper of stars" above Gatsby's bright lights. All of these artists were inspired by the sight of a night sky unblemished by light pollution. Few passages about the powerful "beauty" of the night sky stand out to me more than one by Ralph Waldo Emerson.

> One might think the atmosphere was made transparent with this design, to give man, in the heavenly bodies, the perpetual presence of the sublime. Seen in the streets of cities, how great they are! If the stars should appear one night in a thousand years, how would men believe and adore; and preserve for many generations the remembrance of the city of God which had been shown!

This passage, published in Emerson's essay "Nature" in 1836, always stops me. Just imagine what a starry night would have looked like in Concord, Massachusetts, in the 1830s. In fact, one really has to imagine that dark sky more than a hundred seventy years ago in order to believe Emerson's calling out, "Seen in the streets of cities, how great they are!" To most Americans today, this line would make no sense. In most streets of most cities, the stars are anything but great. One thing Emerson is talking about here is how people take for granted the beauty that surrounds them every day. In order to make his point, he chooses a sight of such grandeur so that readers will truly understand the force of his argument—that we even take for granted something as amazing as the stars as "seen in the streets."

In their many different ways, artists in every medium, inspired by the night, have long asked us to pay attention. All around us, in countless ways, the beauty of the world at night shares itself daily. But because this beauty is as ever-present as the stars were during Emerson's time ("but every night come out these envoys of beauty," he goes on to write), it is too easy to take this beauty for granted.

The spread of light pollution is if nothing else a story of not paying attention, of taking a naturally dark night sky for granted; in the process, we have lost the stars not only from the streets of our cities but from the nights of our lives.

There is a second way I think of night's inspiration, and that comes in terms of "holiness,"—that is, the experience of darkness both literal and metaphorical as a vital part of a human's journey through life. In our world today not only are we addicted to light in the literal sense, we are addicted to light in the metaphorical sense. We live by the slogan "don't worry, be happy" and perhaps too often are quick to reach for pills or drink or the remote control to move us through dark times. But story after story in human history tells us that the experience of darkness is vital to our development as human beings.

In creation myths across cultures, the world (just as in a human's birth) originates in darkness. In these stories, there is no creation without that darkness. In the hero myths of culture after culture, a necessary part of the heroic journey is an experience of darkness. This often takes the form of a literal journey to an under-world, a dark forest, or a cave, often to battle a "dark" creature.

Professor of mythology Joseph Campbell said in *The Power of Myth,* "One thing that comes out in myths, for example, is that at the bottom of the abyss comes the voice of salvation. The black moment is the moment when the real message of transformation is going to come. At the darkest moment comes the light." The message is not that light is "bad" or unimportant but that darkness is an equally necessary part of the story, as necessary as are dark skies in order to see the light of the stars.

We find a similar message in the *Bible.* We are often too quick to assume a basic understanding of light as good and dark as bad (Jesus as light of the world). But in scene after scene—Jacob with the Angel, Jesus in solitude, Jonah in the belly of the whale—darkness is vital, and many of the scenes literally take place at night. In religious traditions as well, darkness has long played an irreplaceable role for the devout believer on a spiritual path. I think of St. John of the Cross in sixteenth-century Spain and his "dark night of the soul." And again, we need to think of how dark the literal night would have been for St. John in order to understand the power of his words. Nowadays, we use that term in a generic way to describe everything from a bad night's sleep to deciding between paper or plastic. The term has lost the weight it once had in a world where everyone knew what a dark night really was.

In short, in literature and art created before the ridiculous spread of light pollution, the darkness of night was a vital part of our most important and fundamental stories. To lose that darkness from our lives is arguably to lose as well the true meaning and wisdom of these stories.

The loss of those old stories, or at least the power these stories once had, may soon be accompanied by the loss of other stories— stories that might not even be created. Henry David Thoreau wrote in 1856, "I should not like to think that some demigod had come before me and picked out some of the best of the stars. I wish to know an entire heaven and an entire earth." Unfortunately, a demigod created by our overuse of light has done exactly that, and

Opposite page: The dark night sky has sparked whimsy, wonder, and imagination for millennia. LAURENT LAVEDER, WWW.PIXHEAVEN.NET

storytellers today are growing up without knowing anything close to "an entire heaven and an entire earth." And I wonder, when you think of all the stories that have been inspired by night—through human history, in literature and painting, drama and dance, anything—what will happen to our creative and cultural life now that generations of Americans (and people all over the world) are growing up without their own stories of the night? What paintings and poems, dances and dramas, essays and novels and photographs are we losing without even knowing it?

And still, I am optimistic. Compared to many of the enormous challenges our society faces today, the problem of light pollution is one we can engage with on an individual, community, and societal level. This engagement will happen because people will spread the word about the true cost of light pollution—to our health, our pocketbooks, our national security, and the ecosystems to which we belong. But this will also happen because of writers and other artists who continue to bear witness, as Henry Beston did eight decades ago, to the "holiness and beauty" of night.

In many ways, as night has long been important for story, now story will be important for night. And everyone has a story about night.

A wide-angle lens was used to capture zodiacal light seen from the Zselic Landscape Protection Area in Hungary. ZOLTÁN KOLLÁTH

I teach at a small college in northern Wisconsin. This year, in support of a recently published collection of essays I edited that testify on behalf of night and darkness, I have been giving a number of readings. Whenever I talk with people about the value of night, no matter where I am, everyone has a story. Some have stories about the night sky, some have stories about the neighbor's light that shines into their bedroom, but everyone has a story. This morning, at a church in Minneapolis, I heard the story of a man who was on a ship on the North Sea, standing on deck alone, the only artificial lights anywhere the ship's red and green running lights and above, a starry sky "unlike anything I had ever seen, or ever seen since."

I heard the story of a woman who grew up on an Iowa farm and remembers the first artificial light in 1948, and how she used to know her directions by knowing the stars but now says she has "a terrible sense of direction; I'm always getting lost." And the man who grew up on a North Dakota farm and used to take his horse out into a field on frozen winter nights to watch the stars. "I never do or see anything like that anymore," he says.

These stories are important. I believe that the more we are able to get people telling their stories about the night, the more people will understand the value of darkness and work to return darkness to our lives.

On my way back into town tonight I drove past my usual exit and kept going for ten minutes to the Maxwell property, an area of wild land donated to the college where I often hike. For the last hour of my drive, a big waning gibbous moon just a day past full led me home, lighting my way. Once I got to this property out in the country, I laced up my boots and tightened my snowshoes. And then for an hour I walked through the woods under a clear sky and a bright moon. At first, my ears hummed from the four-hour drive back from the city, and I kept thinking I saw yard lights through the trees, even though I knew there were no artificial lights close by. As the hum in my ears died down in the still winter night, I realized that the "lights" were ice crystals on bare branches, sparkling in the moonlight. Looking back the way I'd come, the ice crystals in the moonlight appeared as cocoons or spiderwebs woven around every bare limb. I fell into the powdery snow, lying on my back to watch the stars—the Pleiades, Orion, Taurus, the usual winter jewels.

Here was a wonderful experience of night, snowshoeing in the woods outside of town under a big ancient moon. A night I probably will never forget, a story I will tell anyone I meet in the next few days. I'm glad I had somewhere to go to experience this, some way to get there, and the imagination to tell me what I would find when I got there—the twinkling of moonlight on ice, the quiet of the countryside, the memory that will last for years, the experience of what it means to be alive on this world.

Here was the beauty and holiness of night, waiting for its story to be told.

5

HOMEOWNER'S GUIDE TO LANDSCAPE LIGHTING

There are approximately 130 million homes in the United States. In the average household, as much as one-quarter of all energy consumed is for lighting. Light affects the way we live, feel, work, and interact. It is essential to living in our modern society. But a tremendous amount of this energy is wasted because of the ways we use lighting inside and outside our homes.

Nothing can enhance and transform the look of a home like well-designed landscape lighting. The welcoming glow provides great curb appeal and enhances safety, beckoning family and guests and leading them safely from street to entry. Well-designed landscape lighting can also accent the house and yard, indicate direction, and highlight special features. But landscape lighting is an area where more does not mean better. Excess residential landscape lighting has become ubiquitous, going far past the single sconce light at the front door many of us grew up with. Well-designed landscape lighting does not mean the entire property is lit up like an amusement park—no home needs a theme-park installation on the front lawn. The most beautiful, elegant, and functional landscape lighting installations are often the most subtle.

Well-designed landscape lighting gently illuminates traffic areas, showing family members and guests the way to move safely to the house entry. It can softly highlight focal points, special elements in the landscape, and key architectural features of the house. What it should not do is blast you in the face, form jarring dots of light marching up a drive or walkway, shine on neighboring properties, or create dark areas that are effectively "black holes" to the viewer. Poorly designed landscape lighting is worse than no landscape lighting at all—both in terms of aesthetics and safety. And too much or badly designed landscape lighting is light pollution.

With the wide availability of do-it-yourself landscape lighting and the growing number of companies whose sole purpose is to design and install exterior lightscapes, the amount and availability of residential exterior lighting have exploded. Homeowners watch design shows or are inspired by point-of-purchase materials to light up their homes so they resemble noon at midnight. Exterior lighting design companies are usually paid by the fixture, so more fixtures mean more income. All of this can lead to poorly executed or overdone lighting installations around the home.

Neighbors argue over "your light" in "my space." This is light trespass, defined as any outdoor light that falls where it's not intended, needed, or desired. In some neighborhoods, night lighting is so bright that homeowners need heavy window treatments to block it out in order to sleep. Many of us know a neighbor who leaves outdoor lights on all night long, shining into others' bedroom windows. This light encroachment can be thought of as actual trespass on property. Good exterior lighting is directed to shine where it's needed—and not anywhere else.

Of course, it's rarely feasible to turn out all the lights. We need appropriate outdoor lighting to see better at night, and to be safer in our cities, towns, and neighborhoods. But following a few simple guidelines can go a long way toward creating functional and aesthetic outdoor lighting.

Practicing good lighting can be as simple as installing a PARShield on a typical floodlight. SUSAN HARDER

Good lighting design means:
- Avoiding glare, which decreases visibility
- Using the right amount of light—not too much and not too little
- Using fairly uniform lighting so eyes can adjust easily
- Avoiding deep shadows and creating smooth, gentle transitions from light to dark
- Avoiding light trespass; shining light only where it's needed and wanted
- Eliminating uplight (we don't live in the sky!)
- Avoiding the clutter of too many lights
- Saving as much energy as possible

Subtlety, efficiency, and dark-sky compliance—making sure fixtures don't contribute to artificial sky glow—should be the goal of all outdoor residential lighting. Well-designed exterior lighting is an intersection of art and science. It includes the consideration of several variables: lighting task, fixture selection and placement, lamp choice, energy source, amount of light, and timing of light.

Fixture Choices

Form follows function in most good design. Landscape lighting is no exception. Lighting fixtures are now available in a wide variety of styles; it is important to match the fixture to the lighting requirement. To light the way to a front door, for example, floodlights that create glare, light trespass, and energy waste would be a poor choice. A floodlight would light the way, but not without blinding guests and leaving the house hulking in a dark void behind the light source. A better choice would be path lights that gently illuminate the walkway to the door. Just as in interior lighting, the types of fixtures used in landscape lighting should be chosen based on the specific lighting requirement.

Accent Lights. Accent lighting is often used in the landscape to serve a purely decorative purpose: emphasizing special features, such as a unique specimen tree, fountain, or piece of garden art. But it can be functional, too, defining a space by gently lighting fencing at property boundaries or washing down a hedge or planting to bring depth to the night landscape. Accent lighting used in this way can also serve as background light, helping to create uniform lighting without harsh transitions. It can work beautifully to lead guests down a driveway or walk without creating glare or light trespass. Remember that with all accent lighting—and with all lighting, for that matter—a key premise of dark-sky compliance is pointing the fixture down, not toward the sky.

Good lighting should be subtle in order to draw attention to the illumination, not the source.

Pole or Bollard Lights. This style of lighting features a lighting fixture mounted on a post of any height, from one foot to thirty feet. Pole lighting shines down to illuminate anything from a parking area to a residential pathway. Pole lights can contribute to light pollution if they're not properly shielded—the fixture should direct light only where it's needed by adding shielding and adjusting the angle of the beam if possible. Pole lights that are dark-sky compliant have appropriate shielding and can be a very efficient and effective way to light space.

Step or Path Lights. Like the tiny, low lights in movie theaters used to unobtrusively illuminate aisles and stairs during a show, step or path lights are subtle, energy-efficient, effective ways to light exterior walks and stairs. Set in masonry or installed on low fixtures that wash light down and across the steppable surface, these lights increase safety, direct traffic, and can be a beautiful complement to stonework, decking, or landscape installation.

Wall-Wash Lights. Also called soffit lighting, wall washing is a popular technique for illuminating surfaces. Inside or out, wall washing generally uses fixtures installed in a series that wash down the side of a structure to create artistically illuminated walls. This technique can be employed not just on architecture but also on plantings, hedging, and fencing. Like accent lights, wall-wash lights can help create uniform illumination and eliminate dark spots. Remember that wall-wash lights should always direct the light downwards.

Task Lights. Particularly useful for places like exterior kitchens or gathering areas, task lights should illuminate the specific work area and little else. Used in conjunction with accent, path, and wall-wash lighting, they can be an exceptional way to light an outdoor entertaining area. Most task lighting takes the form of recessed lighting, pendant lights, or table or floor lamps. True to its name, task lighting can usually be extinguished when an area is not in use.

Security Lighting. Frequently used to deter intruders or increase the feeling of safety, security lighting in the residential landscape often takes the form of banks of glaring floodlights. Ironically, there is no correlation between increased lighting and decreased criminal activity. A softly and uniformly lit yard will usually provide the visibility necessary to achieve good security, but many people *feel* safer with specific security lights.

If security lighting is essential to create a feeling of safety, it should be designed for maximum efficiency. The most effective security lights are activated by passive infrared sensors. This sensor activation acts as a deterrent, since the potential intruder is instantly aware of being seen, as well as a way to alert the homeowner. The addition of a photocell to the security lighting assembly will ensure that these lights activate only when it is dark, thus saving energy. (Because crime prevention is such a big concern in many landscape lighting plans, it is covered in detail in the next chapter.)

Above: These residential lights will never trespass into the neighbor's window. Directing a light source downward means the light falls where you need it while interiors remain dark.

At right: Glaring floodlights directed outward instead of downward are one of the worst light polluters.

General Lighting. For the overall illumination of an area in a residential landscape, general lighting is used to illuminate parking areas for guests, pools for nighttime swimming, entertainment areas, and porches when they're in use. General lighting can take the form of recessed lighting on porches and in entryways as well as decorative sconces at doorways and on the entries to garage areas and outbuildings. Like pole lights, general lights should be shielded so the light shines only where it is directed.

Floodlights. Floodlights provide illumination across a wide area, typically when high levels of light intensity over broad areas are desired. With the exception of motion-activated security lights, there is little reason to use floodlights at any residence. They are a very poor substitute for good task or ambient lighting; they are unattractive, cause glare, and are notorious for light trespass. Floodlights also tend to deliver a bright cone of light that leaves a very dark edge, making dark areas appear even darker and providing an uneven—and therefore danger-ous—light experience.

Elements of Smart Landscape Lighting

Lamps. Lighting experts define "lamp" as the bulb used in a light fixture. Lamp choice makes a huge difference. Lamps contribute greatly to overall lighting quality and the amount of energy consumed. Incandescent lamps create a warm, traditional glow. Halogen lamps provide a full-spectrum light experience, perfect for a jewelry store and quite lovely in the landscape. Both incandescent and halogen lamps lose up to ninety percent of their energy in heat, however–an incredible waste. For the greatest energy efficiency, LED (light-emitting diode) fixtures are recommended. LED fixtures are low voltage, which provides an immediate energy savings over in-line voltage fixtures. They are also safer in the landscape, since they present virtually no risk of electrocution. LED lamps use a fraction of the energy that a similar incandescent or halogen lamp would use. And the individual lamps can last ten times longer than incandescent alternatives. The latest iterations of LED lamps are also beginning to address concerns about their cool blue color, which is the least favored wavelength for outdoor lighting. They are now available in a variety of colors as well as "warm white" and "golden candlelight" versions designed to mimic the familiar incandescent glow.

Despite much recent hype, compact fluorescent lamps are not the best choice. These lamps threaten to be a disposal problem of tomorrow since every one of them contains a small amount of the hazardous and highly volatile neurotoxin mercury. In addition, CFLs do not work well in cold weather, necessitating a cold-weather ballast. Most detrimental to their being a highly desirable choice for landscape lighting is their inability to be dimmed easily or attached to a motion senor, timer, or other electronic control. Until mercury-free CFLs are developed and widely available, LED lamps (especially those with a correlated color temperature under three thousand Kelvin) will be preferred.

Brightness. As bulb choices become more energy efficient, it is tempting to add more fixtures to a home or yard. This impulse is not only unnecessary, it is counterintuitive. An over-lit exterior can be jarring and distract from the aesthetic of any landscaped area. Lights that wash out the features of the objects they illuminate are too bright. Lighting elements that compete with each other for brightness are never welcome in a well-designed lighting scheme. While brightness levels are widely subjective, a good rule of thumb is to try to comfortably see the object being illuminated from a vantage point in a dimly lit area. If visibility is easy, the existing light is sufficient. If you are forced to squint or blink at an object bathed in light, some lighting should be removed.

Direction. Probably the number-one mistake in landscape lighting is to use fixtures or placement that shines the light upwards. Fixtures should always direct light where it is needed-and that is neither up nor directly into people's eyes. A poorly designed six-foot pole lamp is a perfect example of what *not* to do. These driveway markers shine light directly into faces, causing pupils to shut down, making seeing clearly almost impossible. A better choice are soft lights that wash down a few tree trunks or other landmarks, gently indicating the way, or low lights that splash over the face of a fence or wall along the drive, giving an ambient glow for direction. Even small path lights that spill light across a walk or driveway are much better choices.

Timers and Other Electronic Controls. Another common mistake is to keep outdoor lights burning all night long. An effective method of controlling light pollution is using a simple curfew to turn lights off at an appropriate time. Timers, motion or occupancy sensors, and dimmers should be a part of every landscape lighting design. Dimmers can be installed on nearly every light outside (and inside) the home. Well-designed light-ing installations generally come on just as dusk is falling. They are programmed to turn off a few hours later, or are turned off manually when the household retires for the night. This simple effort can go a long way toward reducing ambient light at night and preventing an enormous waste of energy.

Dimmers save electricity and greatly affect the mood of a space. When the homeowner is expecting guests, the landscape and exterior lights could be at fifty percent. On evenings when a small glow is wanted, they could be dimmed to ten percent. Unlike the rheostats of the past, dimmers no longer reduce the light by dispersing the energy as heat. Dimmers today actually deliver the selected amount of energy to the fixture, saving elec-tricity.

Motion and occupancy sensors have been around for many years. These sensors for outdoors are effective when used for safety and security lighting. They also work well in areas that require long stretches of lighting, turning up the lights just ahead of a person or a vehicle and then automatically turning them down or off a few minutes later. While motion sensors are popularly thought of as security features and are attached to floodlights, they can be used with lower-watt path or task light-ing as well.

Saving Energy. Exterior and landscape lighting provides the perfect opportunity to use low-voltage or even solar-powered fixtures. Because they are so easy to work with, low-voltage fix-tures have particular appeal to the homeowner or designer who is not an electrician. The voltage carried by these systems is so low that the risk of electrocution from cutting a wire or working in a wet location is greatly reduced. A vast selection of low-voltage fixtures is available from lighting sellers, home improve-ment or hardware stores, and neighborhood nurseries. Solar fixtures, too, have come a long way the past few years; both the variety of fixtures available plus the sophistication and efficiency of the solar collectors have been increased. These options are

*High mounting and full shields provide safe, subtle lighting for the path
without drawing attention to the light source.*

great ways to go green in the landscape. Solar-powered choices provide the additional benefit of a low lumen output, which avoids overbrightness.

Evaluating Lighting

Whether you are a homeowner or an exterior lighting designer, how do you evaluate landscape lighting? Asking these questions is a great way to start:

- Does the area need to be lit?
- If so, for what purpose?
- What level of lighting is necessary?
- Do any of the light fixtures emit light above the ninety-degree plane?
- Is there any light trespass onto neighboring properties?
- Is glare—or perceived glare—a problem?

Depending on the answers, here are some basic guidelines:

- If the answer to the first question is no, don't light the area, or turn off the lights when they're not needed.
- Direct light only where it is wanted.
- Eliminate all light emissions above the horizontal (ninety-degree) plane.
- Minimize light output at high angles (seventy to ninety degrees from the vertical).
- Conceal the lamp source and bright reflector sections from direct view.
- Use energy-efficient lighting sources and bulbs.

The International Dark-Sky Association Fixture Seal of Approval program certifies exterior lighting fixtures as dark-sky compliant. In these approved fixtures, the bulbs are never visible, and shielding assures that no light above ninety degrees is emitted. Additionally, light emanation between seventy and ninety degrees is minimized. These fixtures keep the light focused down where it is needed, but only conscientious use of light can create a dark-sky-friendly landscape.

Anyone who lights an exterior space is in a perfect position to help raise awareness and influence lighting in our neighborhoods, towns, and cities. By example, they can educate others on how to use light appropriately in the landscape, save energy and money, and even help reclaim views of the true night sky. By practicing good lighting, we can educate our neighbors as we are educating ourselves, create a pleasant ambience around our offices and homes, and increase demand for lighting design and fixtures that follow dark-sky guidelines. We can work to design and install outdoor lighting that accomplishes these goals, demonstrating a greener approach to lighting the way.

6

CRIME PREVENTION AND SAFETY

There is an old story about a board of directors meeting of a major drill-manufacturing company. One speaker after another extolled the virtues of the company's product. It had the highest sales. It was the drill most in demand. New resources were being diverted to improve the quality of it. Finally, an elderly board member stood up and said, "This company is in desperate trouble and will fail unless everyone understands that customers do not want drills. They want holes! Someone will come up with a way to eliminate the need for holes, and this company will still be making drills. I fear we have missed the point!"

It is common for people to talk about or describe their problems and needs using solution language. So we might say we need more light when what we actually mean is that we have trouble seeing in a dark location. Impediments to visual accessibility might be the real problem. This logic suggests that in the case of lighting, *what people really desire is to see things, not have more light.* When used improperly, lighting itself impedes vision through glare or the veiling of a subject. More light is not necessarily better. This confusion about the objectives of outdoor security lighting directly affects the way many such systems are designed. Too often, these systems feature *too much* lighting, which does little to enhance security and can, in fact, hinder it.

Studies indicate that there is no conclusive correlation between night lighting and crime. Most property crime is committed during the day, or inside lit buildings. Outside illumination can actually draw unwanted attention to a home or facility and help criminals see what they are doing. Lights triggered by motion sensors are much more effective in indicating the presence of an intruder.

Outdoor lighting should provide real security, not just the feeling of safety. Effective security lighting starts with determining and illuminating target areas such as entry points. Using shielded fixtures at these points is beneficial in two ways. First, glare is decreased or eliminated. Uncomfortable or temporarily blinding, a glaring light can distract the eye and cast harsh shadows that create easy concealment for a trespasser. Second, shielded fixtures help a home- or business-owner control both the placement and amount of light used for security. Entrances, windows, and gates can be the focal points of a lighting design that does not overilluminate but allows adequate and uniform visibility that dissipates shadows.

People can see more in soft lighting than they can in spotlights because they can see beyond the point of illumination. Eyes take up to twenty minutes to completely adjust to the dark—longer for aging eyes. Fully shielded lighting provides enough illumination to see the surroundings while reducing excess light that's harmful to night vision.

A vast array of security studies and the professional opinions of public safety experts suggest the need for appropriate outdoor lighting based upon clear behavioral objectives, reasonable uniformity to enhance visual adaptation to varying light levels, control of light trespass or spillover, reduction of glare that blinds the user of the space, and the use of the latest technology that is cost-effective and smart.

Before and after: A 500-watt floodlight (above) and the same light angled fully downwards, showing no glare (below). The adjusted light is unlikely to create a nuisance and increases subject visibility (although it's still brighter than necessary).

Crime Prevention Through Environmental Design (CPTED)

Crime prevention through environmental design (known as CPTED) is a concept that has received considerable interest during the past few decades. It's based on the theory that the proper design and effective use of the built environment can lead to a reduction in the incidence and fear of crime and to an improvement in the quality of life. The concept is supported by the fields of geography, psychology, and criminology, where it has long been known that the design and use of the physical environment affects the behavior of people, which in turn influences the use of space, leading to either an increase or decrease in exposure to crime and loss.

CPTED has been used to reduce crime, premises liability, and fear in a variety of settings: schools, neighborhoods, convenience stores, malls and shopping centers, parking structures, transit sites, hotels, hospitals, office buildings, and parks. State statutes, regulations, and safety standards have been developed to promote the use of CPTED concepts. (It is important to note that CPTED does not replace traditional approaches to crime and loss prevention; it is a tool that helps remove many barriers to social management.)

Traditional crime prevention usually relies almost exclusively on labor-intensive or mechanical approaches (guards, hall monitors, and police patrols are examples of labor-intensive strategies; security cameras, locks, alarms, and fences are examples of mechanical approaches). These methods incur costs that are in addition to the normal requirements for personnel, equipment, and buildings needed to carry out human activities.

The use of CPTED concepts requires that human activities and spaces be designed or used to incorporate natural strategies for crime deterrence. Human beings are born with natural responses to certain environmental stimuli. Other responses are learned within the context of culture, education, training, and experience. Tests with newborn humans and animals reveal that they inherit natural responses to visual stimuli. For instance, when newborns are shown a film in which they seem to approach a cliff, the subjects will automatically react when they think they are going over the edge. Visual stimuli are some of the most important to humans, but by no means are they mutually exclusive of other forms of perception.

Perhaps the best way to understand the impact of visual space is to use a camera as an analogy. Cameras are designed to simulate functions of the human eye. The lens admits light that is reflecting from objects. Objects have altered the light energy, which allows the images to be represented on the film. Similarly, the back of the eye absorbs the light energy and interprets the various wavelengths into three-dimensional shapes. This is where the similarity ends, because the field or width of vision and depth or distance of vision of a camera may be altered—the human eye records the environment with a *fixed* field of vision and depth perception.

Humans establish visual bubbles, which vary in depth, height, and width according to territorial definition and geography. A visual bubble is defined as that space in which a person consciously recognizes things within the environment. Most environmental cues are dealt with subconsciously, outside of the visual bubble, unless something unusual happens to bring one of these elements to a conscious state. For instance, an opaque fence in a person's backyard will establish the outer limit of a visual bubble. Symbolic fences may create the same effect of psychologically obscuring what is happening on the other side, or outside of the visual bubble. Thus, a trespasser who thinks that a solid fence provides concealment unknowingly has the added benefit of the fact that neighbors or passersby are probably not looking anyway.

Why is this important to a study of CPTED? The answer is that the visual sense scans the middle and far environment to collect information for immediate survival and protection. Environmental cues are assessed by all of a person's perceptual systems. But the visual sense provides information about hazards, way-finding, identity, and attractions. For instance, highway safety is almost completely linked to visual perception. Traffic engineers know that certain locations have a high number of accidents, and some locations seem to induce excessive speeds. Noise-reducing barriers along the sides of expressways are known to create tunnel vision in drivers, which can lead to an increase in accidents.

In the hours of darkness, lighting dictates visual perception of an outdoor space. The boundaries of lighting are the boundaries of the visual bubble. Understanding the requirements for good lighting helps define the space and control or influence the behavior that occurs there. Lighting choice consciously and unconsciously affects the perception of safety. But lack of lighting and fear of crime does not correlate to lack of lighting and occurrence of crime.

Perceptions of Safety and Security

Most studies of the impact of outdoor lighting on crime and fear of crime occurred between 1967 and 1979. Since then, studies have centered on the types of outdoor lighting and issues associated with light trespass or spillover. The initial studies uniformly found that lighting positively affected people's perception of safety. The results of large-scale lighting demonstrations were equivocal, however. There were many problems in developing an experimental base for testing the effects of lighting on significant sections of communities. The studies have all identified different requirements for rural, suburban, and urban outdoor lighting. The need for increased lighting levels was directly correlated to the density of population and land uses.

One project conducted studies of outdoor lighting that consistently showed that residents felt safer when outdoor lighting was increased, but there were no consistent effect on reported crimes. One research problem was controlling for increased crime reporting by residents who were more aware of outdoor activities. These and other studies of the time indicated that people were more likely to use outdoor spaces when lighting was increased. And there was no attempt to account for the possible displacement of crime from increased lighting areas to others that were less lighted.

Another study group evaluated forty-one different lighting projects under the U.S. Department of Justice's National Evaluation Program. The overall conclusions were: (1) there was a strong indication that increased lighting, and lighting uniformity, decreased the fear of crime; (2) there was no statistically significant evidence that street lighting affected levels of crime; some projects showed large decreases in crime while others seemed to have no effect on reported levels.

It is interesting to observe that the research community has not conducted many studies of light and crime since 1979. One possible influence has been the consistent field experience associated with uses of outdoor lighting to deal with individual crime sites. Anecdotal and professional expert opinion provide overwhelming support for the use of appropriate and uniform outdoor lighting for streets, pedestrian areas, and open public spaces. Finally, it seems that outdoor lighting consistently produces a feeling of greater safety among the users of space, regardless of the impact on crime.

British department of transit studies of perceptions of crime in transit locations focused on likely users of these spaces. The studies consistently ranked the presence of visible staff ahead of security cameras and lighting as important to users' perceptions of safety. (These studies also found that transit passengers generally felt safer on their way home from the train station than going to the trains.) This research method using surveys of likely passengers is complex but holds great promise for understanding the relation between outdoor lighting and crime.

Glare can decrease uniformity and create perceived pockets of darkness. Look again! The man in the bottom picture has become nearly invisible by relocating to a shadow immediately under the light pole.

The biggest problems in the research of outdoor lighting's effect on crime are the study methods. Comparisons of pre- and post-data are easy to obtain on a specific location. But it is impossible to conduct the same comparative analysis on a large scale, such as a section of a city or region within a state or province. There is no way to turn the lights off and then turn them on again after a year or so of study time. Raw numbers of incidents may be deceiving if more people come to an area that is well lighted because they feel safer. Which measure is more revealing: crimes per number of residents or crimes per number of visitors?

Strategies for Crime Prevention

Identifying behavioral objectives is essential to planning for appropriate outdoor lighting. What do you want to illuminate? What do you not want to illuminate? The failure to develop clear behavioral examples usually results in poor lateral distribution of light, creating spillover into adjacent land uses, with unintended negative consequences. For instance, glare from a brightly lit school campus may be annoying to neighbors, causing them to tightly curtain their windows. They then cannot see what is happening on the campus and so lose any territorial concern they once had for the school, making the campus less secure. This type of light spillover negatively affects the neighbors of parks, public housing, shopping centers, and industrial sites, as well.

After the objectives are identified, specific types of lights and lighting can be used to effectively enhance safety and security.

Sensor Lights. Sensor-controlled lights are a cost-beneficial tool for providing convenient light where and when it's needed. This type of lighting can be effective at preventing crime in a variety of environments: residential, commercial, industrial, and park locations. Criminals tend to avoid sensor lights because the lights draw attention to their presence. Most sensor lights use an inexpensive passive infrared device to activate the light when a moving heat source enters their zone. A timer is used to deactivate the light when the moving heat source leaves. The sensitivity may be set to detect humans and large animals. The sensor lights can contribute to safety in park or greenway areas by drawing attention when people pass though. They offer another benefit, as well: A light that stays on may indicate that a passerby has had an accident and requires assistance.

Sensor lights combined with tiny digital cameras have even more benefits. A one- to two-gigabyte card captures the activity when the sensor detects an intrusion and turns the light on. The cost of these fixtures is usually very reasonable. The card will fit into photo-card slots on most computers and provide images of whatever triggered the light.

Windows that increase surveillance work with downlighting over an accessway to increase security.

Lighting Uniformity. Visual adaptation in all outdoor lighting situations is important for safety. Vision is lost when a person moves from a reasonably lighted place to a dark one. Likewise, vision is lost when one moves from a dark place to one that is reasonably lighted. Problems can occur when a person looks at a reasonably lighted area that is immediately adjacent to a bright one. The higher level of light causes the human eye to perceive the reasonably lighted area as dark, so visual accessibility is impaired on a broad scale, which negatively affects security. Lighting uniformity does not mean that light levels must all be exactly the same, but it requires gradual buildups and reductions to let eyes adapt to the changes.

Lighting strategies can help define movement from public to semi-public to private spaces in order to enhance territoriality. As well, spillover controls on outdoor lighting can be effective in defining the borders of properties. Variations in lighting will help in way-finding and in the perception of security. A buildup of light should be used to accentuate the locations of pedestrian and vehicle entrances to malls and shopping centers, preferred pedestrian pathways in parking structures, and on campuses and parks.

Glare produces sight loss when a person moves from indoor to outdoor areas, and vice versa, at night. A brightly lit mall, movie theater, or gym will reduce vision when patrons go outdoors into much lower light levels. A gradual diminishment of light from primary hallways to entrance foyers to lighted canopy areas outside will improve vision and perceptions of safety. Many malls place brightly lit food courts, arcades, and movie foyers at entrance and exit locations, producing the negative effect of moving too quickly from bright to dark spaces. As well, glare from poorly aimed or unshielded outdoor lights will impede visual access by creating a foglike effect at crosswalks and parking lots. Pedestrian-vehicle conflict will occur, and the veiling effect creates confusion and can hide potential offenders.

Low evenly spaced lighting improves visibility by providing a uniform light source.

Colored Lighting. There is no rule that says all outdoor lighting should be the same color. Some may argue that a common color of outdoor lighting is aesthetically appealing, but these arguments fail to address the need for behavioral objectives for outdoor lighting. Each element within the outdoor environment may have different objectives for behavioral management.

An area may be securely lighted with monochromatic luminaires that are placed uniformly and minimize glare. The loss of visual acuity and color rendition from this type of light can be mitigated by the use of whiter light on monuments, landscape elements, and sensor lights. The human eye processes various wavelengths of light energy in different ways. Accordingly, a lighting plan that has clear behavioral objectives will most likely require a variety of light sources to support the objectives. The illumination of monuments, landscape elements, artworks, and ornamental design features of buildings and parks will attract the attention of passersby, which will increase natural surveillance. Many of these attractions are locations where vandalism and personal crime occur. This type of lighting will reflect on surrounding areas, without the need for additional outdoor lighting.

Reflected Light. Bright, reflective colors absorb less light energy than do dark colors. Brightly painted surfaces in parking structures and building facades or outside walls will be more aesthetically appealing while reflecting more natural light onto adjacent sidewalks and parking areas. Overhead areas in parking structures and under bridges will reflect existing light better, too, if they are brightly painted.

The "wet look" is used in advertising photo shoots to reflect natural light up from paved areas to the subject. The same benefits of reflected light can be obtained by applying a reflective sealer on asphalt parking lots. Bright paint may be used on sidewalks or smaller paved areas, such as under canopies at service stations. Concrete can be used instead of black asphalt to provide better reflection. If used wisely, reflective material provides the opportunity to use much less source lighting. Second- and third-hand light is a wonderful way to use existing ambient light as an alternative to increasing lighting. Glass-wall buildings and reflective vertical surfaces can bounce light twenty-four hours a day. Local design guidelines and lighting ordinances and bylaws should recognize this concept.

Lighting Parking Lots. Landscaped islands are a favorite element used by designers to hide or screen off the mast-mounted lights in parking lots. The tree canopies impede the lighting, however. Crime analysis studies demonstrate that the dark areas of an otherwise uniformly lit space are the places where the most crime occurs. Yet people will park next to islands to get the protection from sunlight only to come back at night to find their car broken into. A smart lighting program will put the luminaries *above and below* the trees and use timers to shut down the upper lighting when the parking area is closed (the lower lights could be extinguished during winter months when screening deciduous trees and bushes lose their leaves). Sensor lights can be very effective in reducing fear and deterring potential offenders.

A "rule of thumb" that has been adopted worldwide is to require complete visual access between tree canopies and shrubbery. The most common rule is a three-foot maximum height for shrubs and bushes and a seven-foot minimum height for the lower canopies of trees (the "3-by-7 rule"). This concept emerged from the science of horticulture to promote healthier relationships among trees and bushes, which compete for water, nutrients in soil, and UV light. It works for security, as well.

Liability Concerns

It is commonly known that an expert witness in security will prevail over a lighting expert when the absence of appropriate light is the crux of a premises liability case. A higher standard is applied to spaces and areas which are held to be inherently dangerous due to the type of land use or to previous victimizations. Special circumstances within a specific location will dictate the need for appropriate light levels. The key to this is the presence of "due diligence" on the part of the property owner or manager to provide a safe environment. Due diligence is defined as attempts to "do all things reasonable" to protect a space or area. A smart lighting plan makes good sense over a plan that is merely trying to save money. The expert witnesses will respond well to a plan that is multifaceted and clearly demonstrates an attempt to do all things reasonable to light a space in a way that deters crime yet also protects the dark night sky.

7

COMMERCIAL, INDUSTRIAL, AND STREET LIGHTING

It is estimated that at least twenty-five percent of all light pollution is created by commercial and industrial lighting. Images of the earth at night clearly show many of the world's great industrial zones, from the petrochemical facilities of the Middle East to the fishing fleets of the Pacific Ocean. The commercial centers of the world are easily recognized by the density of light at night. Sadly, most of this light is not needed and is the result of careless lighting practices and wasteful overlighting.

Industrial light pollution is not often easily seen by the public. Industrial sites are usually inaccessible and removed from population centers. Only occasionally is the impact readily apparent, when industries are surrounded by a community. Examples in North America include the petrochemical industry's processing and refining centers, shipping and container centers, rail yards, and storage lots. Commercial lighting, on the other hand, is part of every community in the world. Parking lots, storefronts, shopping centers, gas stations, and car dealerships fight for attention using excessive lighting so they are conspicuous in the night environment. Adding to the visual noise and glare are lighted signs; some sources consider signs the second largest cause of light pollution.

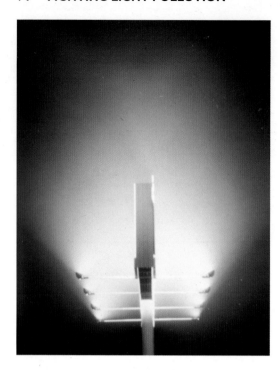

Billboard and sign lighting can be a formidable contributor to a city's total light and should not be overlooked when creating an ordinance.

But it doesn't have to be this way. The environmental impact of commercial and industrial light pollution could be dramatically reduced with responsible design practices, beginning with context-based light and energy-use limits.

Outdoor commercial and industrial lighting is mostly utilitarian and usually consists of box or cylinder luminaires mounted on poles. Such systems are often chosen to meet recommendations of the Illuminating Engineering Society (IES). These recommendations are primarily concerned with providing the proper amount of light for a given task as measured in foot-candles (in the U.S.) or lux (the rest of the world). Since the amount of light varies from point to point, it is generally evaluated by the average light level and uniformity of the lighting over the area. There are hundreds of IES recommendations, for everything from parking lots to soccer fields.

In recent years, the IES (and its international counterpart, the International Commission on Illumination) has begun to address the environmental impact of outdoor lighting. The latest standards divide the world into five different outdoor lighting "zones" so that lighting recommendations can, in theory, be more sensitive to the environmental context. Unfortunately, most IES lighting recommendations are old and assume only one lighting zone—the new recommendations are slow to be developed. This is one of the many reasons that a large percentage of outdoor lighting systems in place today are simply too much for the surrounding environment. But as a primary principle of responsible design, the amount of light should be adjusted to the situation and ambient light condition. In North America, the use of IES Recommended Practice RP-33 to make light-level choices is recommended. Here, the IES clearly states that regardless of the specific foot-candle suggestions it makes, it is the responsibility of the designer to interpret the recommendation within the context and adjust the amount accordingly.

Commercial lighting is regulated by numerous codes and standards. These typically include local planning regulations, safety listings, electrical codes, energy codes, building code emergency egress requirements, and local lighting ordinances. And—whether they are actually needed or not—most commercial lighting systems operate all night, every night. For this reason, energy efficiency and length of lamp life usually dictate what lighting equipment is used. In the U.S., high-pressure sodium (which has a pinkish yellow tone) and metal halide (with a bluish white tone) lamps are most often selected because of their long life and energy efficiency. (While LED lighting systems promise to outshine other lamps in these areas, claims about LEDs have yet to be fully proven. One of the most important possibilities of LED light sources, however, is dynamic lighting—in which the light level can vary or even be turned off when it's not needed.)

Standard Lighting Products

There are five primary groupings of commercial and industrial lighting products:

- Recessed and surface-down lighting, for mounting into or onto canopies, overhangs, and other structures
- Pole-mounted luminaires, for parking lots, driveways, and the general outdoor hardscape
- Decorative and area-lighting luminaires, designed to be attached to buildings
- Low-level lighting, including step lights, bollards, and other types
- Floodlights, for illuminating building facades, sports fields, and general area lighting

There are also a number of other groupings, such as commercial landscape lighting, decorative commercial lighting, and so on, with a more limited impact. Although they will not be addressed here, similar principles of conservation and environmental responsibility apply to them as well.

For each general grouping, there are significant opportunities to employ better practices to get environmentally responsible results.

Recessed Lighting. Downlighting under soffits, overhangs, canopies, and portes cochere is an effective and responsible method as long as the luminaire is *recessed* and the bottom lens *flat and flush with the ceiling.* Sag-lens and drop-lens luminaires create glare with little other benefit and should be avoided. A wide range of light sources can be used, ranging from incandescent and compact fluorescent to LED, high-pressure sodium, and metal halide. Adjustable recessed luminaires can be used to accent specific task areas, such as gas pumps. Gas stations are notorious for being overlit, creating excessive brightness and glare as well as general light pollution. In these cases, excessive light levels are used as a commercial attraction. Signs and other design elements should be used instead to attract attention with less environmental impact.

Pole Lighting. Pole-mounted lighting is mostly utilitarian and intended to have minimal environmental impact. As a general practice, luminaires with no upward light and carefully restricted light between eighty and ninety degrees (almost horizontal) should be used. This mitigates most of the light pollution. Many of these luminaires are simple boxes or drums with flat-bottom lenses. A new generation of LED luminaires with similar characteristics is becoming available, although its mitigation measures might not be so obvious.

Properly designed pole lighting should be tall enough to light mostly downwards. Many communities ban tall poles, with unintended consequences. Properly shielded luminaires on tall poles can light large areas with a minimum of offsite light pollution. Short poles, on the other hand, encourage floodlighting that invariably produces sky glow, light trespass, and glare. The key to controlling offsite impacts with well-shielded pole lighting is setback, the distance between the light pole and the property edge. Setback should be two-and-a-half to three times the pole height from the property line. But many pole luminaires can be equipped with a "house-side shield," allowing the backside of a pole luminaire to be even closer to the property line.

Sometimes pole luminaires have a decorative function. Unfortunately, decorative pole luminaires, such as those with acorn-shaped glass domes, usually cause light pollution. There are several types that are specifically designed to mitigate pollution light, however:

- "Indirect" luminaires, in which uplight is captured by a reflective top cover. The result is softer downward light suited for walkways and other pedestrian areas.
- "Faux" traditional luminaires, in which the real lamp is hidden in the top of the luminaire, and the optical system acts like a flat-lens box. Sometimes there is a false

chimney, looking like an old gas-lamp, where the lamp would traditionally be located.

• Open-bottom luminaires with a bell or dome shape.

For large areas, such as airports, sports complexes, stadiums, rail yards, shipping areas, industrial plants, and petrochemical facilities, there are a number of special high-performance flood-lights that prevent upward light and offsite glare. Many of these luminaires also improve energy efficiency, and this cost savings typically offsets the added expense of using the better equip-ment. Also, in today's market, there are many responsible choices using fluorescent, compact fluorescent, HID, and LED light sources for literally every type of pole luminaire.

Building-mounted Lighting. It is usually cost-effective to mount lights to buildings instead of poles. Such lighting is inex-pensive and appealing to the bottom line. Unfortunately, it often encourages the use of luminaires that are aimed outward, with light pollution as the result. Two types of building-mounted luminaires cause most of the problems: floodlights and wall packs. Both throw light over a wide area. Outfitted with a pow-erful lamp, these fixtures light an area by brute force. Side effects include glare and wasted light directed upward.

There are quite a few inexpensive well-shielded floodlights and wall packs on the market today, and they should be used to avoid the problems caused by unshielded fixtures. At a mini-

A building-mounted flat-lens luminaire.

mum, a luminaire with a cutoff line that does not trespass off the site should be used, although it may still create onsite glare. A better solution would be to limit the lighted area to about three times the mounting height of the luminaire. Obviously, mounting the luminaire high on the wall can light a large area with a minimum of offsite impacts if the luminaire is properly shielded. Two very good solutions include a "full-cutoff wall pack" designed specifically to prevent offsite impact and building-mounted versions of flat-lens pole lights. These luminaires should never be tilted up, however, as this can create the worst kind of offsite impact.

Some commercial lighting sites require decorative lighting, such as ornamental post lights, lanterns, and lighting for signs. Great care in choosing decorative lighting should be exercised. It is usually possible to find a well-shielded decorative luminaire for almost every purpose.

Low-level Lighting. Low-level luminaires include bollards, step lights, marker lights, path lights, and landscape lights. Usually these luminaires employ low-wattage sources, and their proximity to structures minimizes their contribution to pollution. Still, these luminaires can cause offsite impacts, especially glare. Manufacturers offer shielded versions of most of these luminaires—in most cases, the best design solution costs little or no more than a light-polluting one.

Street Lighting

In the twentieth century, lighted streets and roads became an international symbol of civilization and industrial progress. Myths of improved security and safety encouraged the rapid proliferation of lighting with little or no care for the environment. As a result, street lighting now causes more than fifty percent of the world's light pollution. A principal reason for the proliferation of street lighting was to create a profitable off-peak load for electric companies that had excess generating capacity at night. With other important carbon-reducing uses of electricity at night (such as electric cars) becoming viable, it is time to challenge how much street and roadway lighting is really needed.

A case in point is freeway lighting. Even heavily traveled multiple-lane freeways throughout California no longer employ continuous lighting; lighting is limited to ramps. Yet accident statistics are no worse on for these roads than on fully lighted freeways in other parts of the country that have similar traffic. Likewise, continuously lighted streets in neighborhoods have no better safety and crime statistics than comparable streets with no lighting or intersection lighting only. In fact, in a major study in Chicago, significantly increased lighting resulted in more night crime, not less. Experts now agree that in many cases streets with less lighting or even no lighting are as safe as streets where conventional lighting is used.

Many communities and state departments of transportation justify continuous lighting along streets by referring to American Association of State Highway Transportation Officials recommendations. But in fact, AASHTO clearly states that lighting is not always warranted. According to AASHTO, the "warranting" process is part of community planning and should take into account a range of considerations, including traffic flow, community needs, and the environment. Community planning boards and councils often believe safety and security myths because they and the community have never been presented with alternatives. The result is unnecessarily lighted streets, light pollution, and a large electric bill for the community.

Communities should ignore the myths concerning safety and security and invite the participation of police, fire, and other emergency workers, as well as community leaders, as they develop street-lighting plans. An open discussion about lighting throughout the community, keeping in mind the cost and environmental impact, will help the community to identify real and proven needs rather than allowing history to dictate current and future lighting specifications.

Members of the community should work to identify the commonly occurring conditions in the community and thoroughly investigate lighting benefits and drawbacks. They should suggest that lighting planners use the five lighting-zone system and an "overlay zoning" approach to separate land use zoning from zones of environmental sensitivity. They should also produce a lighting zone map. Finally, community planners should review IES RP-33 and develop community lighting standards for each zone. These standards should establish best practices taking into account all available techniques for mitigation of light pollution.

Street Lighting Fixtures. Traditional streetlights feature ornamental enclosures surrounding a light source. In the early days of electricity, three general shapes became particularly popular: globes, acorns, and lanterns. The importance of lighting in the appearance of the streetscape cannot be understated. City leaders often choose traditional luminaires, but, due to their cost, usually relegate them to downtowns, historic districts, and themed developments. Traditional luminaires are in fact among the most inefficient and light-polluting of all lighting systems. Less than twenty percent of the light energy actually illuminates the street along which it is located. More than fifty percent goes skyward, wasting energy and causing all types of light pollution. While it is possible to improve the net efficiency of traditional streetlights using internal louvers, upward light and glare are still problems.

The most widespread streetlight fixture is the ubiquitous cobrahead luminaire, featuring a flared, rounded head that usually houses a drop lens. Developed to be inexpensive, easy to maintain, and efficient, cobrahead luminaires were designed specifically for roadway lighting and are usually mounted twenty to thirty feet above ground. Increased light levels are often created by using high-wattage lamps. Most cobrahead luminaires are mounted to extended mast arms, placing the luminaire over the roadway and moving the pole away from the roadway for

At right: Globe-style fixture.

Above: Acorn-style fixture. SCOTT KARDEL

At left: Lantern-style fixture.

added safety. Like traditional streetlights, cobrahead luminaires send a tremendous amount of light upwards and sideways, where it's not needed. They also cause glare and cast unwanted light great distances.

To avoid these problems, flat-lens cobrahead luminaires have recently been developed. Critics note that photometric differences may require there to be more flat-lens luminaires than drop-lens luminaires to provide the same amount of streetlighting. But dark sky advocates note that flat-lens luminaires prevent uplight and minimize glare. In the best cases, there is little or no performance difference between these luminaires, and flat-lens luminaires should always be employed. In the worst cases, flat-lens cobrahead luminaires may require eight to nine percent more poles, but only in continuous lighting applications.

Energy Concerns. In commercial, industrial, and community lighting, design decisions include significant consideration of the lighting's energy efficiency. After all, most outdoor lighting systems operate almost five thousand hours each year, and each watt of power will consume about five kilowatt hours a year, costing anywhere between ten cents and a dollar, depending on utility rate and source of energy. Two important new classes of luminaires have recently been introduced with energy efficiency in mind: high-performance, dark sky–friendly luminaires in ornamental and traditional shapes, and LED lighting systems in many different shapes, some dark sky friendly and some not. The first class is a minor compromise in several respects. Using the lamp and optical system of a flat-lens, fully shielded luminaire, this class features decorative elements that mimic traditional luminaires but have a slight effect on performance. In general, there is a small amount of uplight caused by the support stems, lens, and decorative elements. There is also a loss of efficiency compared to other options. But compared to traditional luminaires, they are a significant improvement.

LED lighting systems promise superior control of light onto the roadway, preventing waste into the sky and adjacent areas. Some LED luminaires have no uplight, and an amazing eighty percent of their light is in the useful angles for street lighting. Not all LED luminaires direct light this efficiently, however. A lot of decorative LED luminaires perform similarly to decorative luminaires.

Each of the possible light sources for outdoor lighting has advantages and disadvantages. The following chart compares the different light source possibilities. (Costs included here represent 2010 averages and will vary as sources become more or less popular and efficient.)

High-pressure sodium lamps are used extensively throughout North America because of their very low cost and ability to operate in almost all climates. As the efficiency of LEDs

Lamp/source type	Color	Cost	Advantages	Disadvantages
Incandescent		$7.50/million lumen-hours	Warm toned. Low cost of source and control gear. Optically efficient. Instant on/off operation at any temperature. Dimmable.	Very short lamp life and very inefficient.
Fluorescent and Compact Fluorescent		Linear: $1.50/million lumen hours Compact: $2.50/million lumen hours	Choice of color temperature. Low to moderate cost of source and control gear. Instant on/off operation. Moderate to long lamp life. High to very high efficiency. Dimmable.	Not optically efficient. Not suited for very hot or very cold climates. Lamps contain mercury.
Induction		$3.50/million lumen hours	Choice of color temperature. Instant on/off operation. Very long lamp life. High to very high efficiency. Dimmable.	Optically inefficient. Lamps contain mercury. Moderate to high cost of source and control gear.
Metal Halide		250w: $2.50/million lumen hours 39w: $8.00/million lumen hours	Choice of warm or cool color. High efficiency. High optical efficiency. Moderate cost of source and control gear. Wide temperature range. Moderate to long lamp life.	Warm up and restrike time are significant. Lamps contain mercury and other heavy metals. Do not dim well. Expensive when using low wattage lamps.
High-pressure Sodium		250w: $1.50/million lumen hours 50w: $3.50/million lumen hours	Very high efficiency. Good optical efficiency. Low to moderate cost of source and control gear. Wide temperature range. Long lamp life.	Warm up and restrike time are significant. Lamps contain mercury and/or other heavy metals. Lamps contain sodium. Do not dim well.
Low-pressure Sodium		$3.00/million lumen hours	Very high efficiency. Wide temperature range. Very long lamp life. Moderate cost of lamp and control gear.	No color rendering. Warm up and restrike time are significant. Lamps contain sodium. Do not dim well. Not optically efficient.
LED		$5.00/million lumen hours	Very high efficiency. Good in cold and moderate climates. Very long lamp life. Excellent optical efficiency.	High cost of source and control gear. Sensitive to heat. New technology, long-term performance not fully tested. Quality varies among samples.

Color Key

1800K, monochromatic yellow	2200K, yellow-pink, poor color	2700-3000K, warm, good color	3500K, neutral, good color	4100K, cool, good color	5000-7500K, cold and bluish

improves, their cost falls, and their long-term performance is confirmed (especially in hot climates), they are expected to become dominant. In general, fluorescent, LEDs, and induction lamps are best suited for adaptive lighting applications.

Different light sources have somewhat different effects on light pollution. Monochromatic light, preferably in long wavelengths, such as yellow, is preferred by astronomers because it can be filtered out, leaving the rest of the spectrum for observation. Also, it creates less atmospheric scattering, reducing sky glow. Other warm-colored sources are preferred for their attractive appearance and because they create less glare. Conversely, bluish light sources are least preferred as they cause the most atmospheric scattering. In addition, low-wavelength light causes the most glare and has the greatest effect on circadian function. Blue-rich sources can produce improved human visibility at extremely low light levels, however.

It's important to remember that many outdoor lighting applications do not need all-night illumination. Lights can be dimmed or extinguished or change color to reduce environmental impact. In many cases, lights can be turned on by motion sensors or other detectors and turned down or off when they're not needed. In other cases, lights can be dimmed or extinguished after a curfew time. For instance, communities should consider turning down or off neighborhood streetlights after the normal bedtime for children.

An unexpected benefit: Retrofitted streetlights in the foreground drastically reduce glare from the roadway (visible in the background).
R. DECHESNE/ROYAL ASTRONOMICAL SOCIETY OF CANADA

THE IDA FIXTURE SEAL OF APPROVAL PROGRAM

Because choice and availability of well-made fixtures is crucial to improving the nighttime environment, IDA reaches out to the lighting industry to promote luminaire shielding and dark sky–friendly design. In 1988, limited choice in commercial and residential fixtures made it exceedingly difficult to design lighting systems that minimized glare and light trespass. IDA began a dialogue with lighting manufacturers, distributors, and trade groups to encourage fixture shielding.

Some members of the lighting community were receptive to this message and have worked to create quality products that minimize light pollution. As manufacturers began to develop luminaires that control light more effectively, the market for such products started to expand. But the terminology involved with these luminaires was confusing to the layperson. Distinctions such as "full-cutoff" fixtures, which allow no light to shine above the ninety-degree angle, and "cutoff" fixtures, which allow as much as fifteen-percent uplight, were made, and the question of what qualified as a dark sky–friendly fixture arose. The difference between the terms "shielded" and "fully shielded" also led to confusion. IDA sought to remedy this problem by offering third-party approval on fixtures that satisfy an established set of standards.

IDA started the Fixture Seal of Approval program in 2005 to certify qualifying products as "dark sky–friendly fixtures." The program recognizes lighting manufacturers who integrated the concept of full shielding into their fixture design and to encourage market expansion of such designs. Manufacturers of these products create choice within the lighting industry: a choice to adopt smart lighting instead of misdirected light and energy efficiency instead of unnecessary waste. This choice is vital to the ongoing reduction of light pollution.

To earn the seal of approval, manufacturers submit a luminaire's photometric data. An approved fixture must be fully shielded to emit no light above a ninety-degree angle. IDA analyzes a luminaire's Upward Light Output Ratio (the amount of upward flux a fixture produces). If criteria for full shielding are met, the fixture is approved, and the manufacturer receives a certificate and the Fixture Seal of Approval for use on their product lines. The program is open for submissions from any manufacturers worldwide that produce luminaires believed to be dark sky–friendly.

The FSA program has been wildly successful for both IDA and the lighting manufacturers who join. The IDA seal is gaining worldwide recognition and becoming a selling point for manufacturers and vendors alike. The market for dark sky–friendly products is expanding as technology improves and companies work to design sleek, stylish, efficient fixtures. Over one hundred manufacturers have joined the FSA program, creating approximately three hundred approved fixture models. Lamp varieties that can be placed in the fixtures expand the choice of dark sky–friendly outdoor lighting fixtures to more than a thousand.

The seal of approval helps consumers gain access to products that will accomplish the goals of reducing stray light and creating better ambiance. Many of these products come with the added bonus of increased efficiency. People who install dark sky–friendly fixtures outside a home or business are making a conscious choice to respect neighbors, save energy, and balance backyard ecosystems. Because many residential fixtures use new technology that is only beginning to be developed for commercial use, these products can cost more than other choices. Despite the increased efficiency that usually offsets up-front costs over the life of the bulb, approved residential fixtures are not always found in neighborhood stores. For a large selection of residential fixtures, customers can buy from online specialty stores such as Starry Night Lights. Dark sky advocates are encouraged to ask their local hardware and building supply stores to stock FSA dark sky–friendly fixtures.

On the other hand, the market for approved public and industrial lighting has grown exponentially. Manufacturers at industry trade shows display the FSA logo on many of their products. Builders, architects, and city planners who attend these shows can choose from a variety of dark sky–friendly products, including LEDs, induction lighting, and high-pressure and low-pressure sodium fixtures. The market now includes so many qualified products that there is virtually no difference in price between approved and unapproved fixtures.

The availability of fully shielded fixtures is changing the way cities are lit. In 2002, most of the sixty million luminaires used for public roadway and parking area lighting used drop-lens cobrahead fixtures. Less than ten years later, the majority of newly installed or retrofitted highway lighting products use flat-lens, fully shielded fixtures. These widespread changes are making a noticeable difference in municipal bills and dramatically decrease wasted light.

As successful as these voluntary changes are, no factor encourages technology as much as legislation does. In the past years, Italy and Slovenia have passed national laws that require fully shielded luminaires in all street lighting applications. In 2010, IDA approved several elegantly designed fixtures with cutting-edge technology made by Italian and Slovenian companies.

Of course, fixture design is not the only component of good lighting practice. The amount of light used must be considered and dimmed where appropriate. The Installation Seal of Approval, a new aspect of the FSA program, is being designed to scrutinize the overall quality of an entire lighting system.

Promoting alternatives to outdoor lighting applications offers quantum breakthroughs in night sky protection. IDA approves dark sky–friendly devices to recognize lighting accessories such as ballasts or applications such as wireless dimming controls, timers, and even a radar system that can be used in place of continually running signal lights for aircraft. (The latter system features sensors that use ground-based radar technology to detect aircraft. Upon detection, unlit light beacons begin to flash white during daytime and red at night. As a secondary warning, a radio system squawks "powerline, powerline.") These innovations employ creative technology to replace the need for lighting with a system of greater functionality and less light. They are revolutionizing the way the world looks at the necessity of lighting!

Obstacle Collision Avoidance System warning lights flash on when the system detects the proximity of aircraft. These lights cause a change in the pilot's visual landscape to inherently increase detection. When the aircraft has passed, the lights turn off. COURTESY OCAS

8

LIGHT SOURCE
AND WAVELENGTH

Poorly designed outdoor lighting is one of the most conspicuous forms of energy waste. The global call to conserve resources has cities scrambling to replace public lighting with brand new systems. Technology under development for decades has produced a number of options, many with a potential for energy savings.

Many of these new options have never been applied on a broad scale, however, and may have unexpected consequences if widely used for outdoor lighting. In particular, the stronger blue emission produced by some white light sources, such as blue-rich white LEDs, has been shown to have increased negative effects on astronomy and sky glow and has a greater impact on animal behavior and circadian rhythms than other types of light. Widespread installation of any white light sources rich in blue emission is among the largest concerns of the dark sky movement.

Lamp choices made today will affect night lighting for decades. It is imperative that decision-makers understand the consequences—both positive and negative—of lighting choices. Research in visual, environmental, and health sciences suggests that our understanding of the effects of light at night, in particular blue-rich white light, lags behind the development and use of lighting technologies. Furthermore, such light carries a greater impact to the dark nighttime environment and may have unintended consequences for human vision and health. In 2010, IDA released a comprehensive review paper to raise awareness of likely or potential negative consequences of blue-rich white light and to help governments and the industry balance these consequences against the more widely touted benefits.

Light, Vision, and Correlated Color Temperature

Light visible to humans has wavelengths from about 380 to 760 nanometers (nm), with longer wavelengths appearing red and shorter wavelengths appearing blue and violet.

Different lamps have different spectra: A lamp's spectral power distribution is a quantitative measure of the amount of energy (or power) emitted at different wavelengths. The broad spectral characteristics of different lamps are often discernible to the naked eye. "Warm white" sources, such as incandescent bulbs, emit more strongly at the middle and longer (red) wavelengths. "Cool white" sources with a spectral power distribution favoring short wavelengths cast a light that appears harsher and colder to many observers, even though it may approximate the color of daylight.

An LED's spectral power distribution is assessed by correlated color temperature, which is measured in Kelvins (K). An LED with strong emissions in the blue spectrum has a high CCT; strong emissions in the red spectrum produce a low CCT. Therefore, a 5000K LED gives off significantly more blue light than does a 3500K LED.

Much of the recent interest in blue-rich white light sources stems from two factors. The first is the rapid improvement in the efficiency of white LED lighting. LEDs promise to soon surpass current lighting technologies in the ability to produce light for less energy. Further, LEDs produce light in a way that can be more effectively controlled, increasing efficiency of fixtures and allowing light to be delivered precisely to the areas where and when it is needed.

The second factor is more complex, and more controversial. LEDs can be made in nearly every visible color, but the most efficient formulations are rich in blue light because blue wavelengths activate phosphors that provide the other colors necessary for high-quality white light.

Dark-adapted (scotopic) eyes are more sensitive to shorter (bluer) wavelengths than are light-adapted (photopic) eyes. Therefore, all else being equal, a light source producing more blue light will tend to appear brighter to the dark-adapted eye. However, this perceived brightness is only actualized in extremely low-light situations where the eye is fully dark adapted and dependent on scotopic vision.

Some lighting researchers have proposed "correction" factors that allocate extra lumens to cool white sources. This leads to an apparent additional increase in efficiency for blue-rich white LEDs. Those with more blue emissions—the high CCT LEDs—benefit the most from this correction.

Under normal night lighting conditions, however, the eyes are not fully dark adapted. Lit outdoor environments are typically too bright for full dark adaptation, instead stimulating both rods and cones. This dulls the eye's scotopic sensitivity to blue light and reduces the benefits of the extra lumen "correction" for BRWL sources.

Spectra of five common lighting sources

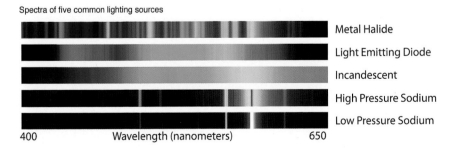

Metal Halide

Light Emitting Diode

Incandescent

High Pressure Sodium

Low Pressure Sodium

400   Wavelength (nanometers)   650

In addition, high amounts of blue light have been linked to numerous concerns:

Glare. Research done as early as 1955 indicates that blue-rich white light causes more glare, with later studies confirming a wavelength of approximately 420 nm to be most closely linked with discomfort glare. It has been widely observed that blue-rich white light headlights on automobiles are perceived as more glaring than conventional halogen headlights. A light source with increased spectral output below 500 nm will increase the perception of glare, particularly for older people, and is more likely to hinder vision than a conventional source of the same intensity.

Adaptation and the Aging Eye. Several studies indicate that blue light reduces pupil size more than other types of light, especially at lower lighting levels. Blue light also increases the time it takes for the eye to adapt to darkness or low-level lighting.

As the eye ages, it requires more light and greater contrast to see well. The lens yellows and becomes less transparent with time, and yellowed lenses absorb more blue light; thus, less total light reaches the retinas of older people, especially when blue-rich white light sources are used. In addition, older pupils are in general more constricted, increasing the amount of time it takes to adjust to different light levels. Blue-rich white light contributes to slowing these adaptation processes.

Atmospheric Scatter and Sky Glow. Increased scattering from blue-rich white light sources leads to fifteen to twenty percent more sky glow detectable by an astronomical instrument than emissions from high-pressure or low-pressure sodium lamps. Because of the eye's increased sensitivity to blue light at very low levels, the visual brightness of sky glow produced by blue-rich white light can appear three to five times brighter than it appears with high-pressure sodium and up to fifteen times as bright when compared to low-pressure sodium.

Blue-rich white light contributes to sky glow in a portion of the spectrum that currently suffers little artificial sky glow. Further, under natural conditions the brightness of the sky in the blue portion of the spectrum is lower. In other words, blue-rich white light introduces a different type of light pollution to a part of the spectrum that is relatively dark and relatively less polluted. Thus, widespread use of blue-rich white light will substantially increase the degradation of visual and astronomical sky quality.

Ecological Considerations. The relationship between artificial light and wildlife has rarely received the level of study that definitively answers questions about spectrum and illumination threshold. While no absolute conclusions have been drawn, previous research suggests blue-rich white light heightens behavior disruption in some species more than other types of light. Loggerhead sea turtles are ten times more likely to be attracted to light at 450 nm than 600 nm, with four Atlantic sea turtle species showing a similar spectral misorientation response.

Light sources that have a strong blue and ultraviolet component are particularly attractive to insects (though broad-spectrum sources are known to attract insects as well). Changes in insect behavior often affect numerous other species that prey on insects, including amphibians and bats. Additionally, the circadian response of wildlife often resembles that of humans; thus even if a species exhibits no behavioral or orientation response to blue-rich white light, such light may be altering the diurnal and nocturnal patterns of wildlife.

Evidence does not indicate that the behavior of all species is altered by short-wavelength light. Some birds have exhibited a stronger attraction to red light, while others avoid it. It is because of these discrepancies and interspecies behavioral unpredictability that IDA advocates for comprehensive studies that investigate spectral effects to be performed before radically different light sources are widely introduced into nocturnal ecosystems.

Circadian Disruption. Light inhibits secretion of the hormone melatonin in mammals. Blue light between 430 and 510 nm shows the greatest disruption of circadian rhythm and melatonin secretion, with peak sensitivity around 460 nm. Some studies suggest that the illumination threshold for melatonin disruption is quite low, but no exact figures have been found. All potential compounding factors have not been ruled out, and crucial research concerning realistic incidental exposure to outdoor lighting, as well as the spectral characteristics of such lighting, has not been published. The effects of blue light on melatonin production, however, and the effects of melatonin on human cancer growth in certain laboratory experiments are uncontroversial. While a firm connection between outdoor lighting and cancer has not yet been established, if true, it is clear that the blue component of such light would be a greater risk factor.

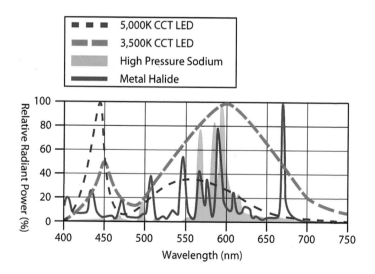

Which Light to Choose?

Objective, professional research is needed to provide manufac-turers with data needed to minimize light pollution while con-sidering other variables. A light source, fixture, or design that optimizes only one aspect will likely fail miserably when intro-duced in a wider context. Sustainable lighting must cross multi-ple realms—technical, environmental, and socioeconomic.

Solid-state lighting has the potential to revolutionize outdoor lighting in a profoundly positive way. LED lighting in particular can be fine-tuned to decrease most negative impact on the noc-turnal environment. The directionality and controllability of such lighting facilitates the large-scale implementation of automatic timers, dimmers, and sensors. LED efficiency and longevity may provide a real contribution to the world's lighting needs.

If developers concentrate on creating high-efficacy lamps rich in warm hues, LED technology could become an outstand-ing source for energy-efficient, night sky–friendly outdoor light-ing. Already, some LED developers are creating a highly efficient product with a spectral power distribution that avoids peaks in any wavelength. Driven by an indicated aesthetic preference for warm-toned hues, these manufacturers are developing high-efficacy commercial LED products with significantly reduced blue spectrum emissions.

Warm white LED technology cannot by itself provide the answer to the world's outdoor lighting needs without the con-tinual application of sensible lighting principles and practices. From a historical perspective, increases in lamp efficiency have not yielded expected savings. Unfortunately, such technological advancements were often used to apply more light, light areas and tasks that were not previously lighted, or make other changes that took advantage of a greater light output for less money spent.

The current initiative to create the most energy-efficient lamp technology threatens to disappoint those who look to

LEDs and other blue-rich white light sources as a technological salvation. While lamp efficiency and consideration of spectral power distribution are significant accomplishments, true night sky–friendly lighting can still only be achieved by vigilant examination of how much light actually needs to be used and implementation of minimum levels required for security and recreation.

Suggestions for City Lighting

Cities ready to make immediate changes to their outdoor lighting may not have time to wait for results of extensive testing of new light sources. IDA offers the following suggestions for communities currently planning lighting retrofits using LEDs:

- Always choose fully shielded fixtures that emit no light upward.
- Use "warm-white" or filtered LEDs (CCT lower than about 3,000K) to minimize blue emission.
- Look for products with dimming capabilities; consider dimming or turning off the lights at late hours.
- Work with utility companies to establish a reduced rate for dimmed or part-time lighting.
- Consider the longevity of the entire fixture over the longevity of the light source alone. Power supply or conduit failure will require fixture maintenance even if there is nothing wrong with the lamp itself.

9

CREATING
PUBLIC POLICY

When there's so much light shining into a limitless sky, it's easy to feel that the problem is beyond control. The truth is that excess artificial light is not only one of the most controllable pollutants facing the world today, it is also one of the easiest for an individual to address. Control of light pollution is a profoundly grassroots issue. To date, almost one hundred percent of lighting ordinances have been achieved by an individual or a small group of committed advocates.

Better lighting is not a zero-sum game. Far from requiring a big budget or creating enemies with deep pockets, smart lighting policy can lower private and commercial energy costs while contributing to overall sustainability. When good policy is enacted, the entire community wins. Since most city leaders have never considered the consequences of artificial light at night, education, perseverance, and good public relations are often the most effective tools.

Ever since the adoption of a ban on outdoor advertising searchlights in Flagstaff, Arizona, in 1958, many communities have sought to limit the detrimental effects of outdoor lighting through civil regulation, and this movement has spread across the United States and the world. Though the losses caused by light pollution were first recognized by poets and naturalists, efforts to limit it through light codes became a movement because of astronomers, the "canaries in the coal mine" of the night. The movement's supporters now span disciplines and cultures, from naturalists and biologists to health professionals, environmentalists, conservationists, neighborhood activists, and nonscientific lovers of the night.

Efforts at curtailing light pollution often focus on the idea and process of drafting and adopting lighting codes. But the problem of light pollution is more fundamental. Though the common impression is that light pollution arises because of what can be called "technical errors" in the use or application of lighting, it is a mistake to focus too early or too heavily on technical solutions. Albert Einstein once said, "We cannot solve the significant problems we have with the level of thinking with which we created them."

The result of a too-narrow focus on technical solutions is that we have many more communities with lighting codes than we have communities where light pollution has been brought under control. This is not necessarily because there are errors in the way the lighting codes are written. It is because the technical errors of outdoor lighting are not the fundamental reason we have light pollution, or even poorly used lighting. We have trimmed the sour blossoms while leaving the roots undisturbed.

This chapter offers a narrative touching on many of the topics crucial to the effective control of light pollution. Because our goal is *stopping and reversing* the degradation of the night environment, we must focus on the entire process rather than the much narrower topic of lighting codes alone.

When someone learns about the generally simple technical errors or oversights in the application of outdoor lighting that lead to light pollution, or how light pollution is detrimental not just to the starry sky but from any perspective, they usually start asking questions: How can such a thing happen? Why would we waste so much light, money, and energy? Why would anyone use light in a way that produces so much glare and compromises visibility so badly? Pondering these questions can lead to insight about the true reasons for light pollution and guide us to more fundamental and effective solutions.

Henry Beston saw the very root of the problem when he asked in *The Outermost House* "are modern folk, perhaps, afraid of night?" In our modern, developed world awash with light, the overwhelming majority would answer "of course we are afraid of night!" If we admit to this childlike fear, we do not hesitate to say that we are afraid of those to whom we believe darkness gives power: those who have bad intent toward us. Addressing this discomfort—this fear—is crucial to addressing the fundamental reasons why we tolerate so much bad lighting and light pollution. We have become convinced that light—in any form, and in the greater amount possible—is the best and easiest way to protect us from these dangers. And there are numerous vested interests that play on this fear, although the idea of light as protection has become so deeply rooted in our culture that it rarely needs encouragement.

The key elements for success in reducing light pollution are not only the technical elements of lighting codes, they are also the educational and procedural elements involved both before and after a lighting code is adopted. Success in executing these non-code elements will make the difference between success and failure in the control of light pollution in any community.

Of course, the code itself must be effective in controlling outdoor lighting while addressing community needs. Some codes, drafted by citizens with input from city officials and lighting engineers, are highly effective in saving energy and money and improving quality of life. However, other dark sky initiatives fail because the proposed code or ordinance does not consider the needs of different elements of the community. Due to the wide variance in size, demographics, climate, and character of towns and cities, an ordinance that works in one area cannot be expected to enjoy the same success in another. Lighting needs also vary from neighborhood to neighborhood. Specifications for parking lot lights in a business district would not help decrease excessive offsite glare from a sports field in a residential area, and curfews for sports lighting would have no effect on obtrusive lighting from a neighbor's yard. For these reasons, laypeople are not always successful at drafting ordinances that control light effectively and are accepted by the community.

IDA has developed a tool that is expected to facilitate the advocacy process for individuals and even to entice cities to initiate outdoor lighting controls from *within* government–propelled by citizen education and grassroots action.

The Model Lighting Ordinance: A New Tool for Cities

In 2011, IDA released a Model Lighting Ordinance (MLO). The MLO, developed cooperatively with the Illuminating Engineering Society (IES), is an outdoor lighting template designed for widespread application in midsize to large cities in North America. The IES's position as a trusted standards organization increases the MLO's credibility with planning officials who already rely on IES recommended practices for outdoor lighting. IDA's input assures that the MLO can be effective in reducing glare and light trespass and curtailing wasted energy.

In the past, lighting ordinances seldom have been initiated by local government. Localities are reluctant to initiate an outdoor lighting ordinance for two main reasons: First, to create legislation from scratch can take a very long time. Second, few within the planning department have any expertise in outdoor lighting and do not have the time or resources to become knowledgeable. Often, a dedicated IDA volunteer spends months or years educating community officials and pushing them to draft an outdoor lighting ordinance to protect the night sky. Local government action is usually driven by the elimination of headaches, where the biggest headaches get the fastest action unless an easy, widely accepted solution is available.

The MLO was designed to vastly increase the rate of ordinance adoption and introduce regulation in the large cities that are responsible for the majority of light pollution. At the time of the MLO's release, about two hundred fifty comprehensive outdoor lighting ordinances have been enacted. Many are in small towns. Control of outdoor lighting systems must be much more widely embraced to make a large impact on overall sky quality.

The MLO is a tool to make this happen. Extensive clauses address numerous potential situations, so guesswork is eliminated. Prescriptive and performance methods offer two avenues of implementation, so governments can custom-design their ordinance to meet their city's specific needs. This approach also ensures enforceability. A built-in user's guide enables lighting novices to understand its language.

These important qualifications may mean that lighting ordinances will start to be initiated by local government. Many environmentally conscious communities will be able to use the MLO to reduce their carbon footprint and meet sustainability goals. Some areas will adopt it to reduce energy consumption to avoid building new power plants in the future. The vast majority, however, will do it to reduce headaches.

MLO Structure and Zones

How can the MLO work in every location? It is no secret that the nighttime lighting of New York City is very different from that of Tucson, Arizona. The two cities are different in almost every respect. As such, their outdoor lighting needs cannot be expected to coincide. Each community possesses singularities that necessitate variations in outdoor lighting policy. To address these widely varied needs in a uniform fashion, the International Commission on Illumination first proposed the use of lighting zones. The zone concept was later picked up by the IES, installed into the California Title 24 Energy Code, and is now a part of the MLO.

The idea is this: Some areas should be protected from all forms of outdoor lighting while in other areas, additional levels of lighting are permitted when appropriate. The MLO recognizes five lighting zones, with the fifth zone reserved for use in only special cases (zone zero indicates an ecologically sensitive area where lighting should be used only rarely). It is recommended that lower-level lighting zones be given preference and that the selection of lighting zones be based not on what is already in place but on what type of lighting is desired. Under this system, communities are able to correlate lighting zones with existing infrastructure but are expected to choose zones that use the minimum amount of light to adequately address an area's lighting needs.

Unlike many ordinances, the MLO limits the total number of lumens per parcel based on the lighting zone. It also moves away from "cutoff" terminology in favor of newer backlight, uplight, and glare (BUG) ratings. The BUG rating for an individual luminaire provides more information as to where it is sending light. Allowable BUG ratings for luminaires depend on the zone in which they are to be installed. The overall intent is to limit uplight, glare, and light trespass whenever possible. The full text of the MLO, including a detailed user's guide, is available on the IDA website.

Tips for Implementing a Light Pollution Policy

While IDA expects the MLO to revolutionize the way cities consider outdoor lighting, implementing a comprehensive lighting ordinance can be a large undertaking. Grassroots advocates are more important than ever in persuading city officials and fellow citizens of the many benefits of adopting sound lighting legislation. Such involvement can also help city government determine the best ways to apply the MLO or other lighting policy. As with any activism, it is essential to grasp the needs—and the fears—of the community, or change will be slow and painful.

Anticipate what best suits the lighting needs of the community. For instance, should a policy call for a curfew for sports lights, or should the lights be shielded, like at this ball field in Flagstaff, Arizona?

The following tips can guide dark sky advocates in lobbying successfully for lighting legislation.

Understand the Area. Essential information can be obtained by studying characteristics of the locality that would be subjected to the lighting policy. How many complaints has the city received for obtrusive lighting? Is there a history of collisions at a particular intersection? What are the listed uses (and hours of use) for a park or public recreation area? These statistics can provide a map of which aspects of an outdoor lighting policy would create the largest benefit to the community. Careful records should be kept of all assessments so that the final policy proposal is defendable—policy seekers must be able to show the reasons behind specific components of the proposed policy if more information is requested.

Be Realistic. Determine, in close collaboration with the jurisdiction, what resources are available for implementing and enforcing a policy. This includes both an evaluation of staff members and time budgets required to administer the policy, as well as the kinds of expertise available in staff or in consultation to the staff. Remember that few communities have room in their budgets to provide staff and time resources sufficient to assure that all the steps necessary to achieve compliance with the lighting code are followed as projects are reviewed, and to rigorously enforce violations that inadvertently or intentionally appear. Yet the availability of resources is not a fixed item: It can be influenced. The more the community is aware of the problem, and the higher the priority assigned to solving the problem, the more resources will become available. In this case, the Model Lighting Ordinance offers a significant advantage, because it can be tailored to the parameters of a community's budget and resources.

EDUCATE! How much of the local population understands the issues associated with light pollution? How many even know what light pollution is? It is of the utmost importance that those seeking light pollution control policy educate fellow citizens. What does it mean for the average person, and what would a reduction mean to them? The IDA offers numerous printable resources for public education. Make this knowledge as accessible as possible.

Build Alliances and Create Stakeholder Identities. There can never be too much support for a policy initiative. Policy-seekers should strive to create a stakeholder identity in light pollution control with all citizens across all segments of the locality's population. Determine the effect of light pollution on local organizations and civic groups, and meet to discuss concerns. Personalize the benefits of better lighting by addressing issues important to the local community: wildlife preservation, nighttime driving safety, health, personal property rights, stargazing, or a multitude of other reasons. What's in it for them? What are they losing because of light pollution?

Gain Publicity and Public Support. Public officials at the local and state level can offer invaluable advice or support.

A community star party is a great way to break the ice.

Investigate the issues each of the officials typically supports in order to know how best to frame your reasons. Seek endorsements from popular public figures. If possible, piggyback with special-interest groups or grassroots organizations that have an established rapport with lawmakers. There is likely to be overlap between dark sky policy objectives and those of other environmental groups.

Establish Credibility. Public policy writing is essentially telling the public what it can and cannot do. It will, therefore, likely always raise at least some opposition. Without verifiable evidence of a problem, the extent of the problem, and its ramifications, it is not realistic to expect support of a policy meant to combat the problem, nor is it reasonable to expect the policy to be enacted. Sound, verifiable, and reproducible science should always be practiced in a quest for public policy, including collecting evidence to show that the problem exists. This includes petitions, written and oral accounts of lighting nuisances, and any other account of unwanted lighting.

Document Evidence and Evaluate Methods. A clear path of reasoning and points considered for each policy component will serve as a good defense for the policy if it's opposed. Quality documentation of local light pollution at the beginning of the policy effort makes it possible to compare a change in light pollution levels after enactment. This evaluation of effectiveness will help defend the policy against attempts to repeal or alter it and will help in correcting problematic components.

Use Specific Methodology. Scientific research is not everyone's specialty. Consider seeking the help of a social scientist who conducts research with the public to help formulate and interpret surveys or other data collection tools that will later be used to support the policy. Researchers and graduate students in social science departments are especially suited to lend expertise.

Anticipate Arguments. Why might people oppose a light pollution reduction policy? Fear of crime is often cited, along with the misconception that policy supporters want to eradicate night lighting. Businesses may be wary that their patronage will decrease without advertisement by brightly lit signs. Never shoot these arguments down, as this will create animosity. Instead, allow stakeholder groups to voice concerns and continue to educate the population.

Do Not Alienate. Realize that many people who currently have offensive lighting may not be aware that the lighting is bothering someone. It may prove fruitful to speak with offenders to make sure they are aware of the problem.

Offer Solutions. Explain how lighting can be modified and where to find appropriate fixtures. Align suggestions with the parameters of the many elements provided by the Model Lighting Ordinance or the proposed code to create the best possible fit with the needs of your community.

ONE ADVOCATE'S PATH

Susan Harder

Our goal is simple: to have better night lighting in our communities. This section is written to help advocates navigate the many ways to help abate light pollution in their communities, based on personal experiences—both successful and unsuccessful.

Remember that success will require change, and change is difficult for many people, even when the benefits are many. King Whitney, Jr., once wrote in the *Wall Street Journal*, "Change has a considerable psychological impact on the human mind. To the fearful it is threatening because it means that things may get worse. To the hopeful it is encouraging because things may get better. To the confident it is inspiring because the challenge exists to make things better."

Dealing with light pollution involves a six-prong approach, and here is my advice and experience with each.

Education

First, educate yourself, continually, on the good reasons for eliminating light pollution and how to do it. Most lighting has been installed without attention to shielding, energy use, bulb type, aiming, intensity, wavelength, and shut-off control, so it's relatively easy to learn how it can be done better. Luckily, high-quality exterior lighting that directs light where it is needed is not rocket science.

Remember that it's helpful to figure out solutions before listing problems. Simply pointing out a problem is not as helpful as recommending a solution at the same time. Self-education about rudimentary lighting design is an essential first step to becoming an effective advocate. Much useful information is available on the websites of dark sky groups such as IDA.

When you have a clear idea of the problems caused by light pollution and the benefits of a sensible lighting plan, you'll be able to educate others. In my experience, most people have not thought about this issue at all, and when they're exposed to the issue, they are eager to see change. Take-home material is always a good accompaniment to formal or informal talks.

Educating others is also a very good way to learn about your community if you have not previously engaged in community activism. Start with those who are willing to listen (but try not to spend all your energy on your closest relatives, because they will tire of it). Of course, educating groups is even more effective than talking to individuals. Form alliances with local groups—join or form a relationship with the group before asking for its help. Civic organizations meet regularly and usually look for speakers on topics of local concern. These groups are exceptionally important because they also represent a voting bloc, and elected officials are accustomed to listening to their concerns and acting on them.

Environmental groups are perfect potential partners in an effort to abate light pollution. There is so much information about the negative effects of night lighting on flora and fauna out there that they will most likely welcome your perspective. They may also help set up night-sky viewings and full moon walks. Many groups have a newsletter, and you can volunteer to educate the editor for an article or write one for submission.

In the early stages of my advocacy efforts, I spoke to local environmental organizations and at Earth Day fairs. Then I tackled the neighborhood associations, Rotary Club, Audubon Society, Sierra Club branch, Lion's Club, electrical contractors, various astronomy clubs, and school classes. Numerous speaking engagements had a snowball effect and led to even more venues. When appearing at events, it's always helps to identify yourself as a dark sky advocate. Wear a tag with your name and "Dark Sky Association/Dark Sky Advocate/Ask Me About Dark Sky Lighting" on it and hand out literature. Pretty soon, whether you like it or not, you'll be considered the local expert.

Local elected officials will usually need to be educated. This can be accomplished by a direct approach (a meeting) or by speaking during the public segment of council meetings. Try to keep it short, simple, and "easy" at first because sometimes these meetings are televised and you will want to appeal to a broad audience. When speakers take up too much time, audiences get impatient and quickly lose interest.

Local civic leaders are easily identified by reading the local newspaper. Contact them to discuss your interests and concerns. It takes some people a bit of time to understand the idea of light pollution and good night lighting. Most, though, will look at the nocturnal landscape differently once they've been introduced to the concept of light pollution. I have found that people will not immediately understand this issue when I first speak with them. In this case, I offer to take them on a night tour. The message sinks in after they've had a chance to see tangible examples. Glare from unshielded floodlights is hard to miss. I try to use municipal floodlights as examples, since I don't want to point out private properties when owners have not had a chance to make changes. Point out good and bad fixtures and the dark shadows that result from glare.

Lead by example: Shield all your outdoor lighting and lead a workshop at a community event to gain support and explain that thoughtful outdoor lighting doesn't have to be difficult to achieve. This inexpensive shield is no less effective for being simply a scrap of gutter.

Most local newspapers will welcome subjects for articles and will likely print your message in a letter to the editor. As you make progress, try to meet with the editor, and you might get an editorial in support of your initiatives. Keep your letters short; brief missives are more likely to get read. Write in a rhetorical style: "Has anyone thought about how exterior lighting causes glare and disrupts our view of the night sky? We can see better with shielded lighting, and more stars are visible." The newspaper will also help you identify local citizens who are trying to deal with community issues. These people might be able to advise you and support your efforts.

Most communities have public-access television stations and local shows so residents can talk about area concerns. Watch the shows and give the host a call when local issues are addressed. Be prepared to answer questions. One question I'm regularly asked is: "Isn't light needed for security?" Remind listeners that lighting alone is not useful to secure a deserted area unless someone is watching the area and able and willing to call for help if they see an intruder.

When talking before a live audience, demonstrations are very effective teaching tools. A simple device that proves very helpful is a clip-on floodlight available at a hardware store. Aim it toward the audience. Then aim it down. Ask them if they can see better without glare. This simple trick used by many successful dark sky advocates, and the consensus is that its effectiveness cannot be overstated.

You may encounter roadblocks at the same place in every speech, such as audience members questioning the security or safety of a revised lighting plan. When you are presenting the same information over and over to a number of new audiences, it can be only too easy for your frustration to become apparent. Try not to be confrontational. Remain educational in your approach. Assume nothing, because even electricians, who are routinely asked to install lighting, may not be aware of how outdoor light behaves or of better alternatives for their customers. Most people will stay engaged if you exercise patience.

Try to reach out to manufacturers' representatives whose products are carried through a local lighting distributor. They will very likely know exactly what dark sky fixtures are and can be very helpful. Reps can also help you find dark sky–friendly fixtures and supply quality lighting plans to an important demographic of builders or homeowners you might not reach otherwise.

I gave a "dark sky award" to a local lighting salesman who educated electricians and selected shielded fixtures to highlight. This is an excellent, low-cost way to generate positive press. All it takes is a frame and a printer (and a photo for the newspaper). Reps will likely still sell unshielded lighting when it's specified, but positive recognition goes a long way in encouraging reps to make recommendations for more shielded fixtures. They know someone is paying attention, and praise never hurts.

Find Examples

It may sound obvious, but first be sure your own home is a good example of quality outdoor lighting. Then you can look throughout your community to find dark sky–friendly fixtures as well as lighting that creates glare, light trespass, or sky glow. Use these examples when you discuss the issue. I mentioned community lighting tours for civic officials—these tours are in fact helpful for anyone you are trying to educate about lighting fixtures because they provide an opportunity to see a variety in use. Examples of good lighting in prominent locations throughout your community will help you draw attention to effective solutions.

To promote an example of good lighting, I donated a Glare-Buster fixture (about seventy-five dollars) to my town, and they put it up on the town hall, gradually replacing all the old wall pack fixtures. It's also an easy and cheap solution to hand out a set of IDA-approved PARshield glare visors to a friendly neighbor with bare floodlight bulbs that can be seen from the road. I developed these because I could not find any other solution to my own neighbors' light trespass into my bedroom. It worked!

LIGHTING BY BRANFORD/THE GLAREBUSTER

An early success I am particularly proud of occurred when I asked a post-top manufacturer with shielded fixtures to lend a sample to replace an unshielded acorn light. The town manager had it installed as an experiment (unfortunately, this was after a pedestrian was killed, possibly due to the glare from high-wattage acorn lights). The next morning, funds were secured and an order was placed to replace one hundred fifty fixtures. This project paid for itself within five years because we only needed a hundred watts per fixture, not two hundred and fifty, which had been the recommendation of someone concerned about safety.

Policies and Guidelines

Municipal facilities do not always go through the planning department or building permit process. If this is the case, it's important to find out which engineering firms are used by the town and try to educate them with a presentation or demonstration. Firms will likely be receptive. Let them know that manufacturers will often provide a free lighting plan for parking lots to conform to recommended guidelines. A lighting designer is preferable for larger projects but not always available or necessary. Municipalities can usually adopt a set of guidelines for outdoor lighting on new construction for their own facilities without legislation. When working with city building officials, repeated contact and interaction is very important. Find out if there will be any new buildings planned and let them know that you'd like to see dark sky–friendly lighting installed. Some officials may not be receptive at first, but remember that the construction is funded by tax dollars, and you have a right to offer input.

Street-lighting officials—if your community has control of its streetlights—will likely be interested in how to deal with all the light trespass complaints they've received over the years. A good approach is to offer a solution, not a complaint. The ubiquitous cobrahead streetlight with a drop lens can be replaced with relative ease if you show your local officials the dark sky–friendly version: a cobrahead with a flat lens. They may even start to order them and use them for replacements.

If your streetlights are handled through a utility, it may be a bigger nut to crack. If you can get a meeting with the utility, use Connecticut's law as an example; it requires full-cutoff streetlights for all new and replacement fixtures, with minor exceptions. The state also has a street-lighting law that applies to the state utility that requires the use of shielded fixtures.

My local electrical utility company provides dusk-to-dawn floodlights attached to utility poles to illuminate private property. These unshielded floodlights are called GE Powerfloods and are common throughout the country. They spread a great deal of light off target and into the night sky and often violate local lighting codes. Since the lights are a major source of light pollution, several dark sky advocates met with employees of the Long Island Power Authority to express their concerns about the floodlighting program. Eventually, this led to a meeting with the chairman, who initiated a complete change in practices, eliminating the unshielded floodlights in favor of full-cutoff mongoose fixtures.

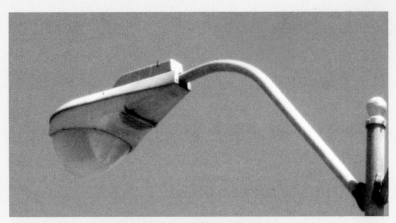

Drop-lens cobrahead luminaire.

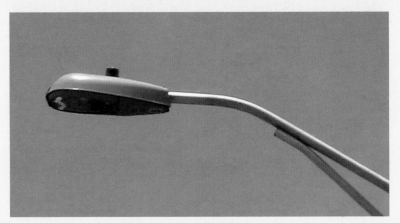

Flat-lens cobrahead luminaire.

One important option for working with utility companies is the creation of a part-night rate for street lighting that dims after a defined curfew, usually 11:00 P.M. or midnight. Cities don't always have an incentive to create public lighting curfews that dim or switch off lights because utility companies may present street-lighting rates in a "dusk-to-dawn" service package. Utility companies may not want to present a reduced rate for part-night service, but it's worth pursuing. Such a system saves money and energy and improves night-sky quality. IDA Northeast Chapter leader Leo Smith created a part-night rate for streetlights in Connecticut in 2010 after three years of working with various public departments. His most effective action was filing a Motion to Intervene with the Department of Public Utility Control, which compelled the utility companies to create a part-night streetlight rate at the time of the next rate increase. New Hampshire is pursuing a similar part-night rate.

Lighting Plan Reviews

As you get more involved in changing the lighting in your community, you will find yourself working with city officials in the various departments that regulate or install outdoor lighting. Find out which agency reviews new site plans for commercial and residential construction. It has probably received complaints if bad lighting has been installed, and these can be a matter of public record. Sometimes a planning department can, by law, require an applicant to submit a lighting plan before it receives a building permit. Offering to help review a lighting plan is the first and best way to be helpful.

If you can, get building and planning departments to accept and distribute an outdoor lighting guide. To help provide a visual guide for my local building department, I created a "Diagram of Acceptable and Unacceptable Fixtures," and the planning department received "Guidelines for Good Exterior Lighting Plans" to hand out. These guides can be an easy way for commercial and municipal facilities to aid building planners with outdoor lighting choices.

The planning department and planning board can usually require better lighting for new commercial sites without legislation. Sometimes each section of a town has a specific planner assigned. Try to meet with at least one of them, as well as with the building inspector and town planner. Learn how to execute and review a good lighting plan.

Try to take a look at a lighting plan submission for an upcoming site. It should be available prior to construction. I have found a great resource for improving lighting plans to be lighting manufacturers. When given guidelines, many manufacturers will supply sensible, dark sky–friendly lighting plans to commercial sites at no charge. Of course, they will only specify the use of their own fixtures, but the property-owners will choose the design.

Regulations

Many people think that just passing a lighting code is the goal. In my experience, a lighting code is only one aspect in addressing light pollution—but it is an important one. I recommend first checking through your zoning code for any references to "lighting" and see what is already there. I found that my own community had a very strict outdoor lighting code but it was not enforced because the terminology was unclear and technically incorrect. It also lacked definitions, which are essential to creating a working code. It read, "The source of illumination shall not be visible across property lines."

Unfortunately, the building inspector (the municipal officer who enforces zoning codes) decided that the face of PAR (parabolic aluminized reflector) floodlight bulbs covered up the "source," which he decided was the filament. I filed a code complaint about my neighbor's PAR floodlights only to discover that the code was inadequate. I brought up the problem at a town board meeting and received encouragement to help them develop a new lighting code. Before I attempted to get a lighting law enacted, I enlisted the help of local organizations to help tighten language and close loopholes. We now have a very comprehensive lighting code (and my neighbor's light has been shielded).

For the most part, community leaders will welcome the help you can offer to enact a better lighting code, since zoning changes are usually initiated by residents. Most towns already have sections in their zoning codes that refer to night lighting in order to control bad lighting (usually called "obtrusive" or "intrusive" lighting).

Each community has its own geographic and social character, personalities, community events, and methods to institute change. If you can talk with someone who has worked with your town government to help enact legislation, this person will be your best mentor. I called upon my local Nature Conservancy and asked one

Perth, Australia, before and after Earth Hour. These photos were taken using identical photographic exposures to accurately show the reduced lighting. JOHN GOLDSMITH/CELESTIAL VISIONS, WWW.TWANIGHT.ORG

of their officers for advice about how to go about enacting a better lighting code.

Of course, enacting a code is an arduous process, but you'll gain allies along the way. I've found that, for the most part, local officials are very interested in improving their communities by regulating outdoor lighting. Many had previously been on the local planning commission and are aware of reviews of lighting plans but didn't know ways to regulate outdoor lighting effectively. You will be surprised at the enthusiasm of many town officials.

You will need a primary sponsor on the council for any legislation. Research council members to figure out who that person should be (a good candidate may have already enacted energy codes or worked on public safety or community character issues) and meet with them. City officials can be very busy, so be polite, but persistent. When you get face time, bring a few handouts about "good" night lighting and the issue of light pollution along with a couple of pertinent articles. If there is a lighting code that has been adopted that you think would suit your community, bring it and discuss it with the official. Make sure the code is borrowed from a community similar to the one you live in. Officials like to know that what they need to do has already been successful elsewhere.

You may be asked to write or submit a lighting code for consideration. All towns are different in size, character, and desire to control night lighting. Determining elements of a lighting code is a big responsibility and should be undertaken only when you have assembled a politically balanced support group. Many municipalities form ad hoc committees to help decide, enact, and provide support for legislation. In some towns, "Dark Sky Advisory Committees" have been formed, and astronomers are usually asked to join. This issue is uniquely bipartisan, so party politics almost never come into play.

Enforcement and Follow-up

Even with a very good lighting code, a formal process of review, an oversight committee, and dedicated town officials, reducing light pollution is an ongoing affair. The people who will have to enforce your lighting code will need to be educated, not just about what the code entails, but how to suggest solutions to violators.

It's important to remember throughout the process that homeowners do not like to feel that their rights are being infringed upon. Community education is an essential tool to avoid or minimize resistance to code enforcement. One way to encourage enforcement is to show people in your community how lighting affects safety. The local police may help enforce a lighting code if they believe that glare is an obstacle to driver safety. Glare can cause accidents, and the police are often acutely aware of this. You might even be able to convince police that it's better if a business-owner turns off lights when the business is closed because then if they see a light on, they will know to investigate.

You will likely encounter allies with the proverbial fire in the belly who can be counted on to show up to speak in favor of changes, write letters to the paper and officials, and file code complaints. These allies can make or break the success of a lighting ordinance. Try to get email addresses from these people and keep them informed of your efforts. Suggest how they can help the campaign and involve them in ways that use their strengths. Remember that good lighting and a pleasant night environment cut across all strata of society: social, economic, political, and vocational.

When local organizations have been introduced to the value of this issue, you can often ask that they add "light pollution abatement" to their stated goals. Many of these organizations have websites and newsletters and can be convinced to add a column or a link to the IDA website. You will likely be asked to write up the text. Keep it simple and in the format of other news items.

Local organizations can be helpful monetarily, too. One of the organizations that asked me to speak offered an honorarium, which I requested be donated to IDA. You could also use monies received to help you distribute printed information or keep for incidental expenses.

When citizens can take pride in their communities they will watch over it more carefully. Lighting awards or recognition can be a great source of pride. Businesses that receive lighting awards are more likely to become involved with the cause. Environmental distinctions for good lighting can also be sought. If a community demonstrates a serious commitment to lighting, it may consider applying to become an International Dark Sky Community for the prestigious distinction it affords.

Before and after: Shielding for lights on the City of Flagstaff fire station creates lighting that is more than adequate and less distracting.
C. LUGINBUHL

Change is possible, but rarely smooth. Be prepared for your community to ask you for more and more assistance as you become known as someone who knows about lighting. Take your time and ask for help or you will be overwhelmed. At first, however, you will be knocking on doors, and some may be slammed in your face. Your education efforts will sometimes be resented, or people may be suspicious of your motives. This is an unfortunate human reaction to change and will most likely be encountered time and time again. This reaction can discourage you, but try to keep your chin up and persevere. There are about ninety-nine people who will quietly support your efforts for every one person who is a vocal naysayer.

And the naysayer might simply not understand the issue. Make an effort to reach out to those who may be speaking against lighting controls to be sure they have been fully educated about the problem. They may continue to disagree, but perhaps less vociferously. Many of the negative remarks center on "security" and "safety." These are concerns that are easily addressed but must be done so repeatedly, and arguments must be made very clearly.

Another source of frustration is the time it takes to complete the political process. It can take years to get even a moderate ordinance passed. You may be shuffled from zoning committees to planning committees before you make a significant contact within the city government. But as you become known, so will your cause. As you gain credibility within the structure of a city or state government, your issue will be listened to—and echoed—by a growing number of people. Patience is the first and best rule, however, when trying to create a change within any government.

Finally, don't seek or expect thanks for all your work, and try to give away the credit. As advocates, we don't change things, we only initiate the process of change. There is little in the way of glory, reward, remuneration, or recognition for this work beyond the satisfaction of making your community a better place to live and work. Keep track of your successes and remember that the changes you help initiate may be slow in coming or may not be on a grand scale but can go a long way. Eventually, you will see the fruits of your labors when lights are installed that conform to good-lighting principles.

And if you are serious about community advocacy, I recommend you treat yourself to a trip to the annual IDA meeting. Nothing is more informative or energizing than being among like minds and sharing experiences. Approaches may differ, but you will learn the most from others' experiences and get new perspective on your own approaches.

LIGHTS OUT TORONTO!

Toronto, Canada, was one of the first cities in the world to make a concerted effort to reduce the incidence of bird collisions caused by bright lights on its skyscrapers. The results of the effort illustrate both its successes and shortcomings.

In the mid-1990s, the Fatal Light Awareness Program, based in Toronto, teamed with World Wildlife Fund Canada to develop the Bird-Friendly Building Program, designed to educate building-owners, managers, and tenants about the hazard that night lighting poses for migrating birds and encourage them in their light-reduction efforts. More than a hundred owners and managers in the greater Toronto area signed on to the program. A few were able to achieve darkened buildings quickly; the majority, however, moved forward at a snail's pace. Participants faced a number of challenges.

Major tenants such as law or accounting firms often worked around the clock in offices on the periphery of a tower. Many of these spaces contained numerous windows and no blinds. As well, many office towers had systems that illuminated entire rooms or floors, not individual workstations. Reducing light enough to help migrating birds was difficult if not impossible under these circumstances.

Another challenge participants faced was the perception that a brightly lit city was more exciting for tourists than a dark one. Vanity lighting to illuminate building exteriors was (and still is) a common architectural practice. Also, many people just feel safer at night in lit-up areas, and some owners were simply not concerned about protecting birds. Problems would arise from a mixture of these elements. New buildings would mushroom within and outside the downtown core, and lighting these shiny new structures was a source of pride for owners, managers, and tenants.

Although the Bird-Friendly Building Program lasted only a couple of years, FLAP was able to document a minor decrease in light levels. Did this result in fewer bird deaths? Probably, but estimates were difficult to make. Toronto's CN Tower (for many years the world's tallest free-standing structure) was the one building that afforded FLAP the opportunity to confirm that the program made a difference. With lights turned off as requested, collisions at the tower stopped except for very rare instances.

In the past decade, FLAP concentrated on educating architects, developers, designers, builders, and city leaders about light pollution's role in bird collision deaths. Their big breakthrough came during a meeting with two environmentally minded Toronto city councilors at which the councilors suggested presenting a Notice of Motion entitled "Prevention of Needless Deaths of Thousands of Migratory Birds per Year in the City of Toronto" to the council. The motion was adopted, and the planning department was directed to produce a staff motion that specified that for all new construction "the needs of migratory birds be incorporated into the Site Plan Review process with respect to facilities for lighting, including floodlighting, glass, and other bird-friendly design features." Early in 2006, this motion was passed unanimously.

A few months after the motion's passage, in partnership with FLAP and a host of other environmental groups and industry associations, the city launched Lights Out Toronto!, a public awareness campaign designed to save money, energy, and the lives of birds. The city published a brochure aimed at sensitizing residential and corporate communities to the hazards that migratory birds face in the urban environment. Ads proclaiming the lights-out message appeared in Toronto's public transit system. FLAP's contact information appeared on both the ads and brochure, linking the program directly with members of the development and architectural communities eager to forestall bird strikes at their proposed buildings. The councilors would also bring FLAP staff into meetings with

A rehabilitated ovenbird is released from an unwaxed paper bag, where rescued birds are temporarily housed. J.P. MOCZULSKI/FLAP

developers, allowing the organization to present the potential problem and propose solutions up front.

Joe D'Abramo, Toronto's Manager, Environmental Policy Section, acknowledged the challenge inherent in implementing a bird-protection policy. "Cities are not organized to deal with such issues," he said. "But, as we find out more about the consequences of our actions, we have to do something about it." FLAP's executive director, Michael Mesure, added that, "Difficult though it may be to measure the effectiveness of this campaign, it's sure to provide added incentive to those companies thinking about investing in energy-efficient measures and techniques."

Bird-Friendly Development Guidelines

Bird-Friendly Development Guidelines were developed and incorporated into Toronto's Green Development Standard in 2007. These guidelines were "intended to provide a list of design-based development strategies available to developers, building managers and owners, architects, landscape architects, urban designers, and planners," addressing key lighting as well as daytime collision issues.

Under the guidelines, bright lights such as spotlights and search-lights that are broadcast over large areas, or other types of event lighting, must be turned off during migration seasons. Decorative lighting of external building features should also be extinguished. Eliminating all but rooftop strobe lights on tall buildings (necessary for airplane safety) and street-level lights enables night-migrating birds to pass over urban areas unharmed. Bright lights high in the sky inexorably attract migratory birds.

The CN Tower was a case in point. When the tower was under construction in the 1960s, powerful lights were used to illuminate it. So many birds were crashing into the half-finished structure that there were rumors of workers bringing out wheelbarrows full of the tiny, feathered corpses. Public pressure convinced management to extinguish the lights in spring and fall, and it followed this routine for many years. But in 2007, the tower installed a multi-million-dollar, energy-efficient LED lighting system that produced a light show at night. This spectacle, meant to attract more tourism to Toronto, and to the tower itself, had the potential to draw huge numbers of migrating birds to their deaths. Luckily, tower managers responded to a FLAP request that they reduce their standard lighting program while birds were moving through the area. The only light visible around the tower during migration now is a streak of light that travels from top to bottom and changes color each circuit, from blue to green to yellow to orange to white. The "top of the hour" lighting effect features a five-minute rotating white light around the Radome (a structure below the observation decks). These measures have reduced—if not eliminated—the bird-collision problem at that site.

Of course, smaller buildings can be a huge problem, too. Fixed outdoor lighting that projects light upwards or increases spill light, glare, and artificial sky glow can be confusing to nocturnal migrants. Almost all night-migrating birds are active during the day but undertake their continent-long journeys at night to avoid predators and take advantage of cooler temperatures. Like us, they gravitate toward the light. Shielded lighting fixtures that effectively project light downwards guarantee birds safe passage.

Advocates note that even interior lighting should be function specific, with individual light fixtures for each workstation; it should not be broadcast far and wide. Lighting that escapes through windows can act as a magnet for birds and contribute to light pollution. The Bird-Friendly Development Guidelines encourage turning off all unnecessary lights at night. If this is not feasible, window blinds or curtains can be used to keep a building dark. Motion-sensitive lighting is a good solution since it will come on only when it's needed.

By no means are buildings the only sites of nocturnal kills. Albert Manville, a wildlife biologist with the U.S. Fish and Wildlife Service, has found that bridge lighting can negatively affect birds. He offers specific recommendations for bridge lighting: "Where pilot warning/obstruction lighting is not an issue, low-intensity, lower-wavelength blue, turquoise, or green lights can be used. They tend not to disrupt the magnetic orientation of the several bird families studied. Red and yellow lights should be avoided." Manville makes reference to blue jelly-jar LED lights. These could be used on suspension cables and bridge decks. Not only do they use less energy, they produce bright but directional light visible from far away while minimizing light pollution that leads to bird collisions.

Birds are more vulnerable to collisions right before they have to cross a lake or other large body of water, particularly when the weather is bad. They seek the closest landfall, and where there are few trees, bright lights may attract them to their doom. Toronto was

Toronto dims its lights during bird migration season. Despite the reduction in lighting, notice how rooftop logos still produce the majority of glare seen in the water. KENNETH HERDY/FLAP

a good example of this phenomenon, with the waterfront only a couple of blocks away from a major thoroughfare and the heart of downtown. Not only were the lit office towers attractive, illuminated billboards along the highway were a potential draw, too. Toronto's council adopted a bylaw regulating the illumination of new signs on buildings and property. It stipulates that from the hours of eleven P.M. to seven A.M., no sign shall be illuminated except those directly associated with businesses that operate during that period, those located in specific downtown business districts, and those erected by charitable, cultural, or community organizations for the purpose of advertising events held by those organizations.

The council also passed a motion making the Toronto Green Standard, which incorporates bird-friendly guidelines, mandatory as of January 2010. It "requires that bird-friendly elements be incorporated into almost all new development" in the city. Toronto has become the first city in North America to adopt such comprehensive bird-protection policies, but New York and Chicago are moving forward with their own programs and strategies: New York has initiated Project Safe Flight while Chicago has started Chicago Bird Collision Monitors.

Starting to Soar

In November 2009, Toronto held the Symposium on Bird Conservation in Urban Areas, drawing on the expertise of organizations such as FLAP and attracting bird-lovers from Toronto, Minnesota, Chicago, Denver, Baltimore, Detroit, and other cities. Most of these activists had persuaded their governments to address the lighting issue. For example, the Michigan House of Representatives introduced legislation mandating that all state-owned or state-leased buildings be required to turn their lights out at night. In Denver, officials agreed to publish a request to turn interior building lights off at night. Only fourteen skyscrapers complied at first, but mayor John Hickenlopper made a declaration in support of Safe Skies Colorado, and the numbers are expected to increase.

FLAP is proud of the progress, but there is now more cause for concern than ever. Mesure puts it bluntly, "In my two decades of bird rescue, I have witnessed a slow but steady decline in migratory bird populations unfold before me. Our spring 2010 bird-collision records serve as a reminder of this disturbing trend. Where we would normally recover over twelve hundred birds in a single spring season, we will likely see fewer than nine hundred for this past migration. I wish I could announce that this decline is a result of fewer lights left on at night, but the fact is that unless we collectively take action to reverse this alarming trend, migratory bird numbers will continue to plummet."

10

NIGHT SKY CONSERVATION

The last half of the twentieth century saw unprecedented natural resource and environmental protection in the United States. A shift toward an analysis of our surrounding environment, our history, and our role in maintaining it permeated the world beginning in the late 1960s. The Clean Water Act, Clean Air Act, National Historic Preservation Act, Endangered Species Act, and, perhaps most impressively, the National Environmental Policy Act, became keystones of the environmental movement. But even in the face of ample evidence that the night sky is an essential component of healthy ecosystems, the Environmental Protection Agency and other federal legislative bodies have not yet taken a significant interest in protecting it through environmental policy. Widespread legislative protection of the night sky as a natural resource requires a sustained effort and high volumes of documented evidence of artificial light's impact on wildlife. IDA works with ecologists to compile this data and consistently updates the EPA on the evolving scientific research, but meaningful policy can take time to implement.

Protection can also be achieved by a shift in culture. Around the world, people are making choices that reflect a new environmental consciousness. "Green" has become a buzzword of the new millennium as individuals (and businesses) strive to conserve the earth's resources: water through rainwater harvesting, fossil fuels by driving hybrid vehicles, energy by using shielded, energy-efficient lightbulbs—simple moves with straightforward results. These behavioral shifts begin with increased awareness and voluntary action and can end in local legislative changes that cement this cultural trend. (For instance, many cities have enacted laws that enforce recycling of appropriate materials.) IDA is applying this mentality to restoring the natural night.

First, though, dark sky advocates must overcome an immediate challenge. This barrier can be thought of as a lack of experience, based on what people *don't* see when they inspect a light-polluted sky. In addition to scientific and environmental consequences, the loss of visible stars means the loss of eons of inspiration, wonder, and curiosity. A clear velvet sky in which stars sparkle like jewels has an intrinsic value. One can hardly gaze at such a vista without pondering the origin of planets and star systems, the outer edges of the universe, or the relative tininess of a life on earth. A dark sky provides invaluable perspective—and many populations will never experience it.

Already lost over urban areas, good night visibility is increasingly vulnerable to degradation in even the most remote locations. Unfortunately, continued degradation has a spiral effect on cultural awareness: the less people are able to experience a striking night sky, the more they have no idea what they are missing. The benefits of personal experience cannot be quantified. Already, millions of city dwellers must drive for hours to experience the night as it once was. If someone lacks the means to leave the city, they may never experience starlight. Worse, if someone has no idea what a dark night holds, why would they want to see it? And if they don't want to see it, why would they take an interest in its protection?

For this reason, night sky conservation programs that protect sites where dark skies are experienced and appreciated are perhaps the most important goal of the dark sky movement.

The success of smart lighting efforts is apparent in residential areas of Calgary, which show a greatly reduced number of visible lights. BUSINESS UNIT IN THE TRANSPORTATION DEPARTMENT FOR THE CITY OF CALGARY

International Dark Sky Places and Other Conservation Initiatives

Effective dark sky conservation takes a dual approach. Outreach programs that raise interest in the night sky work in conjunction with protective legislation and exemplary outdoor lighting. An initiative must embrace the night sky as a natural resource, then act to preserve the integrity of that resource through strictly enforced lighting limits.

Recognizing a deficit in comprehensive tools and techniques for such programs, IDA, together with global collaborators, developed the conservation initiative International Dark Sky Places (IDSPlaces) in 2001. The IDSPlaces program safeguards pristine sky areas (and restores areas with degraded sky views) by requiring steps in night sky protection to be implemented not only by traditional administration but throughout the surrounding community or area. Successful IDSPlaces work with a wide variety of stakeholder groups (including local chambers of commerce, historical societies, astronomy clubs, and environmental groups) to engage the entire area in the act—and the benefits— of conservation.

IDSPlaces build a culture. When one group creates a stake in the night sky, the program can encourage collaboration between public and private sectors. Citizens can inspire park managers and community administrators to protect many of the locations where stargazing is still possible, and inspire private households and businesses to voluntarily control their lighting. Protecting the night sky is not an onerous task—it's one that can be accomplished through simple preservation techniques and sustainable planning foresight. The path to a successful IDSPlace starts with adherence to the principle that should light shine only where needed, when needed, for the minimum time necessary.

The IDSPlace designation recognizes communities and managed land areas that show superior stewardship of the night sky. Areas that receive a dark sky designation not only join an elite handful of parks, they gain a prestigious, globally recognized certification of excellence in preservation and protection of the natural night. Designees distinguish themselves as leaders in ecological preservation and energy reduction and gain leverage in advocating for quality lighting and nightscape restoration in surrounding communities. This leverage is important for increasing protection of the night and builds interest in the idea of sustainable tourism, which not only renews its environment but can be an important revenue source.

IDSPlaces have three main designation categories: Parks, Communities, and Reserves. The Park designation applies to managed public land areas. Communities work with incorporated townships. Reserves protect areas around observatories or astronomically important sites through a buffer-zone system. The execution is slightly different, but all three designations require two main elements: a strict, sustainable lighting management plan that requires use of fully shielded fixtures and low total lumen limits and public outreach that raises awareness on the value of darkness. An IDSPlace designation also requires excellent sky quality relative to the population it serves. Gold, silver, or bronze tier status is awarded depending on sky quality.

IDA also created the Dark Sky Development of Distinction award to recognize unincorporated communities for excellence in light pollution control and outreach.

The idea of special recognition for night sky conservation continues to gain popularity. Groups are creating their own dark sky designations with similar criteria. The Royal Astronomical Society of Canada has a comprehensive program for designation of Dark Sky Preserves throughout Canada, and the United Nations Educational, Scientific, and Cultural Organization's (UNESCO) Starlight Reserves concept, similar to IDSPlaces, establishes criteria for lighting and community education. These criteria not only create protection from the physical encroachment of nearby lighting but celebrate the vital role starlight has played in human culture, history, and scientific development.

Before and after: As part of Mont Megantic's IDSReserve designation, the Village of La Patrie in Quebec reduced unshielded lighting, an action that is now a point of pride for the city. ASTROLAB'S MONT-MEGANTIC NATIONAL PARK, GUILLAUME POULIN

Dark Sky Allies

Land management bodies all over the world are independently recognizing the benefits of preserving the natural night sky (as well as the benefits of energy-efficient lighting), and working to foster dark sky cultures of their own. The expanses of undeveloped land conserved by the National Park Service make them natural dark sky oases. A national park often provides the best place for stargazing for hundreds of miles. Recognizing a special need for managing this resource, the National Park Service formed the Night Sky Team in 1999. This small team monitors night sky degradation, manages park lighting practices, and investigates the effects of artificial lighting on parks. The team is also involved in outreach; park officials hold "park after dark" stargazing sessions to increase awareness and appreciation for the night sky as a recreational resource. Many sessions on local nocturnal wildlife feature the natural night as an important environmental condition.

These efforts are crucial to increasing awareness because they present something that is now very rare: a night so dark that stars and cosmic dust literally illuminate the world. Few people can imagine the objects in a sky this clear. Fewer still will experience anything like it. This on-site dark sky advocacy is powerfully effective because it offers a tangible glimpse of the night in its splendor and, framing the possibility of its loss in that context, creates a visceral need to preserve it.

IDA works with the Night Sky Team on numerous projects and was proud to designate Natural Bridges National Park in Utah as the first IDSPark in 2007. In 2011, IDA signed a cooperative agreement with the National Park Service to evaluate lighting in nine national parks and to recommend dark sky–friendly changes. IDA intends to use this information to create "best practice" guidelines for widespread application in state and national parks, as well as other managed land areas. In addition, groups like Dark Skies Awareness (Dark Skies

Arches National Park in Utah is working to become a Starlight Reserve under UNESCO's World Heritage Sites. DAN AND CINDY DURISCOE

Rangers); Astronomers Without Borders, through their conservation outreach effort One Star at a Time; and the Astronomical League all have conservation-oriented programming. Several initiatives enable schoolchildren to experience a dark night sky and teach the importance of dark sky advocacy. These citizen initiatives are vital to restoring interest in the night sky among the general population.

The Value of Darkness

These programs validate dark sky preservation as a component of mainstream environmental protection. This type of recognition opens the door to considering the dark night sky in two important ways: as a *resource to be preserved* and a *rarity to be enjoyed*. Both ideas can be cited as important reasons for conservation when listing the main benefits of night sky preservation.

One obvious benefit is the support of professional astronomy. Excellent-quality astronomical sites are rare and getting rarer. In addition to having a sky view unpolluted by stray light, certain atmospheric and meteorologic conditions must be met for an observatory site to be viable. Optimal sites are located at high altitudes with little air turbulence and low aerosol content. They must maintain stable, subtropical anticyclone conditions.

This unique combination of circumstances calls for a very specific type of location, usually found on the west coasts of land masses or oceanic islands—places that are very attractive to human settlement as well. Conservation programs emphasize this delicate balance between earth and sky quality for observation purposes. A powerful benefit of parks and tourist areas is preservation of the night sky for astronomical observations. The conservation efforts of Flagstaff, Arizona, the first IDSCommunity, have helped the city retain its viability as an astronomical site. The region houses the U.S. Naval Observatory's Flagstaff Station, the National Undergraduate Research Observatory, and the Navy Prototype Optical Interferometer. The Lowell Observatory, started from a telescope built in 1894, gained international fame as the facility from which Pluto was discovered in 1930. Flagstaff's superb dark sky conservation not only preserves that important link with history, but has enabled the continuing astronomical viability of the facility despite population growth. Lowell Observatory's Discovery Channel Telescope is under construction forty miles away.

Conservation of wildlife offers another invaluable benefit. Sensible lighting applications save energy and money. Benefits to biotica, especially in ecologically sensitive areas, have not been heavily explored to date, but the ecological benefits to nocturnal creatures are undeniable. Dark sky areas that are shown to aid endangered species offer a powerful benefit to wild creatures and the people who love them.

Hortobágy National Park in Hungary earned its IDSPark designation largely to help protect its integrity as the most important bird corridor in central Europe. Hundreds of thousands of gray geese, cranes, and ruffs travel through Hortobágy every year, and rare species of plover, the pale harrier, and the critically endangered slender-billed curlew find refuge in its alkaline marshes. Lighting controls assure that these birds will remain unmolested by artificial light.

Perhaps the most untapped reason to preserve a dark sky is to explore the potential for amateur astronomy and stargazing. The possibilities for these activities are nearly limitless and incorporate a strong cultural and human heritage value into the idea of dark sky conservation.

Before it unveiled the Starlight Reserves concept, UNESCO formed the Starlight Initiative to include night sky conservation in their World Heritage Sites designation, a program that seeks to encourage the identification, protection, and preservation of cultural and natural heritage around the world. In 2007, Starlight Initiative members created the Declaration in Defence of the Night Sky and the Right to Starlight, stating that "an unpolluted night sky that allows the enjoyment and contemplation of the firmament should be considered an inalienable right of humankind equivalent to all other environmental, social, and cultural rights." This view highlights the fact that a so-far-unscathed right is now in serious danger, and its degradation will lead to the irremediable loss of an extensive associated cultural, scientific, scenic, and natural heritage.

Through UNESCO, the Starlight Initiative celebrates the vital role starlight has played in human culture, history, and scientific development. Cipriano Marin, secretary general of the UNESCO Centre of the Canary Islands, Spain, says, "The concept of Starlight Reserve . . . aims to recover and identify the existing values related to the night sky, including those related to landscape, nature, opportunities for science, and, in general, with the associated tangible and intangible cultural heritage."

While closely tied to the culture of astronomy, stargazing activities need not be overtly scientific. "Star lore" and legends relating to the starry sky are becoming popular attractions. In addition, these activities offer insight into other cultures, and can be an important peace-building endeavor. The global astronomy collective Astronomers Without Borders was founded to promote peace across cultures through sharing the borderless sky.

Not least, low-key stargazing events revive a meaningful personal connection with the cosmos.

Economic Possibilities

More and more, dark sky–designated sites are appearing on the growing list of ecological tourism destinations. The idea behind ecological tourism is to nourish a region's natural features–flora, fauna, or cultural heritage–to create tourism that fosters sustainable ecological practices and bolsters the local economy. While dark sky designations often arise from sites with a stake in astronomy, the value of a dark sky as a tourist feature is rising steadily as truly dark sites disappear. The gold tier IDSPark status of Cherry Springs State Park in northern Pennsylvania contributes to its recognition as one of the last, best stargazing areas in the eastern United States. Galloway Forest Park in Scotland has seen a significant increase in its number of visitors since its IDSPark designation in 2009.

Currently, astronomy and dark sky status are closely connected, and astronomical tourism is considered a hot ticket only in relatively small circles. But the increasing rarity of spectacular starscapes and the persistent outreach and education efforts undertaken by sites that receive an IDSPlace distinction are attracting the attention of individuals outside the astronomy community. Increasingly, casual stargazers and weekend campers seek out areas known to have high-quality night skies. This slowly reawakening fascination with the night offers important monetary rewards for conservationists and serious astronomical sites alike.

Opposite page: Cherry Springs State Park in Pennsylvania offers some of the best stargazing in the eastern United States. DAVE WYMER

11

CREATING AN INTERNATIONAL DARK SKY PLACE

The Big Bend Region of West Texas enjoys some of the best night sky viewing in the continental United States. This resource is maintained by the organized efforts of regional advocates who recognize the rarity and value of truly dark night skies, despite the population growth and community indifference that threatened the darkness for several decades.

The Big Bend Region is named for a loop in the Rio Grande, which separates Texas and Mexico. This vast region encompasses over twelve thousand square miles and seven million acres. Here is the high, rugged Chihuahuan Desert, which extends across much of northern Mexico into Texas and covers several mountain ranges. The landscape features volcanoes, canyons, sand dunes, colorful eroded badlands, faults, fossilized dinosaur and marine life, hot springs, old mines, archaeological sites, and a high and dry wilderness ecosystem.

Big Bend has long been a hub of nighttime activity, and west of the Pecos River, the night skies have seemingly always been an iconic element of the landscape, a tradition that dates back centuries to the people who lived there long before Europeans arrived. The indigenous peoples of Big Bend used the sun, moon, and stars as guides to planting crops, hunting game, and navigating. Skies played a significant role in Big Bend native culture, well documented in pictograph and petroglyph rock art. The region's mysterious "Marfa Lights" were explained by the Apache Indians as stars dropping to earth.

More recently, Big Bend's nighttime vistas have been the subject of scrutiny by astronomers at McDonald Observatory, forty miles north of Alpine, atop Mount Locke, one of the darkest major observatory sites in North America. The first big telescope at McDonald was dedicated in May 1939; at that time, it was the second largest in the world.

Far from the lights of larger cities and the clouds of coastal Texas, the observatory has numerous research telescopes and state-of-the-art instrumentation for imaging and spectroscopy. With abundant ambient darkness and a 6,800-foot elevation, the observatory is also in the unique position of being the most southerly major research observatory in the continental U.S., allowing astronomers to view stars and galaxies far into the southern hemisphere more than three hundred nights a year.

Light illuminates the center of this pictograph in Paint Rock at noon on the winter solstice. BILL YEATES

These spectacular dark skies have enticed thousands of amateur and professional astronomers to the annual Texas Star Party, a gathering in the Davis Mountains near the observatory. The event has become so popular that it now ranks as one of the largest gatherings of astronomers in North America.

Eighty miles south of Alpine is Big Bend National Park. According to the National Park Service, Big Bend is the park with the least light pollution in the lower forty-eight states. On clear and moonless nights, more than two thousand stars, planets, and "shooting stars" are visible to the unaided eye. According to Dan Duriscoe of the NPS Night Sky Team, "Several factors make Big Bend an excellent place for night sky-watching. The absence of light pollution, so prevalent near developed areas, makes this area unique among most viewing areas. Big Bend's infrequent cloud cover and low humidity, especially in winter, allow for sharp visual acuity."

Understanding the importance of the night skies in the Big Bend Region makes it easy to see why light pollution needed to be addressed. McDonald Observatory hosts more than sixty thousand visitors each year, providing a direct economic benefit to local communities to the tune of $13.5 million annually. Three hundred thousand annual visitors to Big Bend National Park make protecting the environment, including the preservation of the night sky, vital to the local and regional economy. In fact, tourism is one of the region's primary economic drivers, generating over $210 million annually and responsible for at least twenty percent of the employment in the region.

Battling the Encroaching Threat

Alpine, the economic hub of the Big Bend Region, is located between the McDonald Observatory and Big Bend National Park. Although the sky glow from Alpine is relatively minor compared to larger cities, its proximity to the observatory means that any escaped light over the town has an immediate negative impact. The glow from Midland-Odessa, El Paso, and other urban areas hundreds of miles away is also impeding the ability of McDonald scientists to see into deep space. Stargazers say disappearing stars are akin to the extinction of a species and liken the night skies to other natural wonders that need to be preserved. Astronomers who cast their telescopic gaze over the night sky are increasingly affected by the glare and sky glow from artificial lights, collecting "noisier" data. Some stellar objects can no longer be seen at all.

For decades, the McDonald Observatory has spent time and money working with nearby communities to pass ordinances that require outdoor lights to be directed downward. Six West Texas counties and three cities passed ordinances to control outdoor lighting. However, many ordinances were poorly enforced,

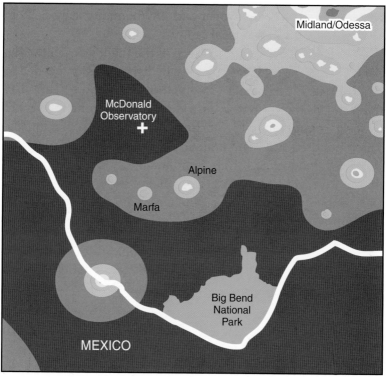

A high-resolution satellite map of the Big Bend region shows light pollution encroaching from Midland/Odessa. BILL WREN

and light pollution continued to increase throughout the region. McDonald astronomers were not really considered a part of the surrounding communities; they were, literally, living in a tower and concerned about an issue that few had any interest in. It was imperative to convince Big Bend residents that dark skies were worth protecting.

One of the great successes of the Big Bend Region's revived dark sky movement is the public concern and sweeping changes in attitude that have developed among residents in the towns of Alpine, Marfa, and Van Horn. Through smart advocacy and community collaboration, a steadily growing number of local residents, business and civic leaders, astronomers, and visitors are realizing they all have a stake in keeping the night skies dark.

Every movement is more effective when it sets measurable goals. In this case, one motivating goal was the idea of earning a number of IDSPlace designations to ensure continued protection for the region's night sky.

Such an endeavor cannot be accomplished without the creation of a pervasive dark sky culture. Astronomers and astronomy-related groups based in Alpine reached out to local businesses to foment public support. The idea of protecting the night sky resonated with regional leaders, and a coordinated campaign began. In 2008, a group of citizens launched the Big Bend Dark Skies website to further increase public awareness. One development decided to earn the IDSPlace Development of Distinction award. This decisive action created the stepping stone for a host of impressive action that is likely to protect the inky skies over the Big Bend region for generations to come.

Sierra la Rana is a family-owned 3,100-acre residential development located two miles south of Alpine. It is part of an 11,600-acre cattle ranch subdivided into parcels ranging from 10 to 200 acres. In 2009, Sierra la Rana was designated an IDA Dark Sky Development of Distinction, becoming the second development in the country to receive this award. The mission of Sierra la Rana is "to preserve the dark skies for the residents of Sierra la

Rana, the McDonald Observatory, and the Big Bend Region today and for future generations."

As part of the public outreach for the designation, Sierra la Rana installed a common astronomy area at an elevation of 5,100 feet with telescope pads (including electricity to each) and plans to install a roll-roof observatory open to residents and guests and to the general public during star parties and upon request.

Sierra la Rana's official recognition solidified the region's dark sky movement in a number of ways. One way was that it helped track economic benefits. Such data can be used to bolster education, outreach, and legislation in the entire Big Bend Region. Since its inception, Sierra la Rana has tracked reasons why prospective buyers have been attracted to the area. Overwhelmingly, prospective buyers have cited the natural beauty, dark skies, wildlife, hiking, and photography as the primary reasons. A year after it was designated, Sierra la Rana was contacted by more than twenty-five hundred prospective buyers. Of the total requests for information, seventeen percent indicated that astronomy and dark skies were the reason they selected Sierra la Rana. In the fourth quarter of 2009, seventy-five percent of all sales were astronomy related, a direct result of the dark skies award and marketing the Big Bend dark skies to amateur astronomers.

Increased Regional Interest

Sierra la Rana management, as well as members of the surrounding community, realized the significance of its recognition for exceptional night sky conservation. David L. Lambert, director of the McDonald Observatory, expressed appreciation, saying, "The work and the standards to which the owners, developers, and future residents of Sierra la Rana are committing themselves will not only help maintain McDonald Observatory's dark skies, they will provide a needed example to the

Residents and visitors use the telescope pads at a star party in Sierra la Rana, Texas. GIL BARTEE

rest of Alpine and the region, where light encroachment is an issue that requires continual attention."

Dave Oesper, moderator of the Yahoo Dark Skies Group, writes of the impact of the regional dark sky action: "As a light pollution activist for the past twenty-five years, I wanted to move to an area that not only has a unpolluted night sky, but existing light pollution laws and a substantial (and growing) population that values the night sky and aims to keep it that way. And the more stakeholders there are to protect and improve the natural nighttime environment, the better. Not only are there many astronomy enthusiasts in the Big Bend Region of West Texas, thanks to a mild climate, abundant clear skies, and a latitude farther south than all of Arizona and New Mexico, but the McDonald Observatory, a major research facility, is in our region, which also strengthens the impetus to preserve and protect our night sky resource here. Add to that the developing

The Milky Way shines in the night sky above the McDonald Observatory.
FRANK CIANCIOLO

astronomy-friendly community Sierra la Rana near Alpine, and this all adds up to one heck of a lot of synergy!"

The success of one development is now being leveraged into motivation for collective adherence to dark sky principles. In the past, efforts to control light pollution were disparate. Now the movement is making tremendous strides toward the uniform protection of the night sky. In 2009, Pete Smyke, a member of a coalition of dark sky advocates and chairman of the Alpine Environmental Advisory Board, appointed Gil Bartee, vice president of development for Sierra la Rana and a member of the Alpine Chamber of Commerce Board of Directors, as dark sky advisor. The EAB proposed seeking IDSCommunity designation for Alpine to the Alpine City Council, and the motion was unanimously approved. A proposed Dark Sky Month was also unanimously approved. The proclamation read, "The month of October has been designated as Alpine Dark Sky Month to educate and inform citizens, businesses, and property owners regarding the use of safe, efficient, and appropriate lighting, with the goal of preserving the treasure of Alpine's night sky, and working toward being designated as a Dark Sky Community by the IDA." Community events during the month of October included dark sky presentations at area schools and the Sul Ross State University Astrotheatre and Planetarium and star parties at Sierra la Rana and Luz de Estrella Winery in Marfa.

Although light pollution has a multitude of negative side effects, the approach of area advocates is to concentrate on the economic impact. Public officials are often most receptive to economic figures, and the conclusion that reducing light pollution will save money and reverse environmental damage helped the ordinances pass.

Dark sky advocates are now extending their focus to include the possible increase in astronomy-based ecotourism. Sierra la Rana's rise in desirability since becoming an official "dark sky destination" is tangible motivation to extend the ethos of dark sky protection to surrounding areas. Alpine and Sierra la Rana are already making the most of the distinction by creating cottage industries related to astronomy, which leads to greater distinction as a noteworthy astronomy region and continues to positively and sustainably feed back on itself.

Alpine is uniquely positioned between two internationally known dark sky sites. With the added benefit of city services, restaurants, and hotels, Alpine is positioned to take full advantage of astronomy-based ecotourism in the region. The verifiable positive economic impact to the City of Alpine led to proposed revisions to the existing lighting ordinance, which include limits on lumens, grandfather limitations on existing noncompliant lighting fixtures, and shielding of fixtures over eighteen hundred lumens.

Development of astronomical ecotourism offers the chance to restructure the already thriving tourism economy as well as attract stargazers and amateur astronomers to a permanent location in the area. In 2010, the city of Alpine unanimously passed an ordinance requiring full shielding of outdoor lighting. The

City of Van Horn passed an identical ordinance soon after, and similar ordinances are pending in Marfa and Lajitas. In 2011, Texas passed a state law banning unshielded lighting within 57 miles of McDonald Observatory. This legislation was sponsored by Rep Pete Fallego of Alpine. Such legislative commitments will further reshape and culture and invoke pride.

Accomplishments

Recognizing the overall economic benefits of good lighting practices, the Big Bend region has set up several innovative grant programs to offset the initial cost of retrofits. In a unique blend of revenue, monies come partially from federal sources, business ventures, the City of Alpine, and private citizens. Through the efforts of the EAB, recommendations by the Alpine city manager, and the unanimous approval by the city council, Alpine has allocated Energy Efficiency and Conservation Block Grant funding to replace 241 open-faced streetlights with properly shielded, energy efficient fixtures to further reduce light pollution and reduce energy consumption by fifty percent.

In addition, a Dark Sky Fund has been established by the Big Bend Astronomical Society to help citizens retrofit obtrusive lighting on private residences. The fund is named after James T. Walker, a dark sky advocate responsible for the passage of the Alpine Lighting Ordinance in 2000. The Alpine City Council has allocated additional funding to the James T. Walker Dark Sky Fund.

The McDonald Observatory and Big Bend Region have benefited greatly from the Al Bowen Light Pollution Reduction Fund established by Texas Star Party participants. With donated funds and the support of local residents, officials, and public utilities, dozens of full-cutoff streetlights and more than four hundred shields have been installed in counties surrounding the observatory.

In a collaborative effort with the National Park Service and a Best Lighting Practices grant with Musco Lighting (administered through the National Park Foundation and the Denver Service Center of the National Park Service), Big Bend National Park completed lighting upgrades at the Chisos Basin in part of their bid to become an IDSPark. (The work was also funded by a grant from the Friends of BBNP and American Recovery and Reinvestment Act funds, with labor and other donations from Forever Resorts, Inc. The project is among the first of its kind in the National Park System.)

Changes to Big Bend National Park

Big Bend National Park is retrofitting one hundred percent of its exterior lighting to apply for designation as an IDSPark. At the time of writing, BBNP's improvements are almost complete. Retrofits were carried out in four phases, one of which was the Chisos Basin, which replaced existing exterior lighting, including building and path lighting, with new LED fixtures. These fixtures use technology that provides comparable light using dramatically less electricity, have an average life of fifty thousand hours, and provide a natural moon-glow color. The fixtures are designed to complement existing infrastructure and blend into the landscape. They are appropriately shielded, providing illumination where needed below the horizontal plane. The fixtures use less than one watt each and replace sixty-watt incandescent and fluorescent lamps. The annual energy costs of lighting for the equipment that was replaced will be $164 per light with the new system–*compared to $3,293 for the old system.* The entire Chisos Basin phase resulted in a *ninety-eight percent reduction* in wattage, energy consumption, and greenhouse emissions.

Lisa Turecek of the NPS states that from a natural-resource perspective, night skies are one of Big Bend's most significant assets. Moreover, the lighting redesign has greatly improved both the quality of night skies at the park and the visitor experience. From a cultural-resource perspective, the installed fixtures are architecturally compatible with architecture and landscape. The expected dramatic reduction in electrical consumption offers a significant cost savings. Lastly, the new lighting provides *better* illumination. Visitor and employee comments indicate that path lighting is actually better after the changes. In fact, the feedback to date—from park visitors and local residents—is overwhelmingly positive. Light that was visible from Terlingua can no longer be seen. Previously, the park lighting cast a shadow on Casa Grande. That has been eliminated.

Cementing a Culture

Collaborative efforts of citizens and regents have created an extraordinary sense of purpose and preservation. Enthusiastic ongoing efforts reflect this mindset, and citizen enthusiasm continues to increase. Recent projects include:

- McDonald Observatory and the Big Bend Astronomical Society continually make presentations to civic groups and governing bodies.
- Through private donation, the McDonald Observatory has purchased hardware to replace and retrofit over a thousand offensive lights with dark sky friendly–fixtures and Hubble Sky Shields.
- The observatory visitors' center secured funds from a private donor to produce a two-minute video on outdoor lighting.
- A demonstration of energy-efficient LED streetlights was sponsored by the city of Alpine. There is an ongoing dialogue with designers, engineers, and manufacturers regarding the need to avoid the use of blue-white LEDs and instead use LEDs that emit a color closer to yellow-orange.
- A Big Bend/Trans Pecos Region chapter of the IDA has been proposed under the guidance of local developers, entrepreneurs, and dark sky advocates Gil Bartee and Dave Oesper.

- Rayford Ball, astronomy professor at Sul Ross State University, has recently rejuvenated the astronomy program at the school and plans to replace the old observatory with a new one at a darker on-campus site.

Through these initiatives, led by a mix of professional and amateur astronomers, civil leaders, and businesses, responsible lighting has become a matter of civic pride. This momentum facilitates even more positive change. And the accomplishments continue. Alpine is more than halfway through the replacement of over six hundred unshielded, visually disruptive streetlights with seventy-watt, full-cutoff, high-pressure sodium fixtures. More than three hundred of the first replacements were made courtesy of American Electric Power at no cost to the city. Revisions to the Alpine Lighting Ordinance have been made to meet IDSCommunity requirements.

The Alpine Environmental Advisory Board has been instrumental in promoting energy conservation and has embraced cost-efficient, dark sky–friendly lighting as a mission that extends to residences and businesses. In 2010, the code enforcement officer in Alpine requested the assistance of the city's dark sky advocates to review the plan of a proposed Dollar General store, resulting in full shielding and a reduction in the number of exterior lighting fixtures.

The movement is catching on in the surrounding regions, as well. The Reeves County Detention Center, near Pecos, is renovating part of its facilities after several costly riots. Some twelve hundred poorly aimed high-wattage floodlights will be removed in favor of high-mast downlighting. Negotiations with the city of Van Horn continue regarding options for mitigating the poor lighting resulting from approximately 150 high-wattage acorn fixtures installed a decade ago. Van Horn officials refer to the two-mile stretch through town as the "runway" and are willing to explore alternatives.

Perhaps the most telling example is the swift action taken by community members when the dark sky sensibility is under

Before and after: Retrofits at a Panther Junction gas station replace 900 watts with 64 watts on high mode or 8 watts on low mode. Light is focused, and the pumps are still clearly visible. BIG BEND NATIONAL PARK

threat. Early in 2010, a new Dollar General Store opened in Marfa, some thirty-five miles from the McDonald Observatory, with what most agreed was highly offensive outdoor lighting. The *Big Bend Sentinel* published a front-page story and editorial cartoon addressing the issue. Prior to construction, the general contractor was never informed of any regulations. Several Marfa residents contacted dark sky advocates in Alpine and at the McDonald Observatory. In March 2010, the Marfa City Council approved recommendations by the City Compliance Officer and Bill Wren of the McDonald Observatory, working with Dollar General Management to bring the store into compliance. Steps taken included turning off roof-mounted flood lights permanently, turning off internally lit signs at closing, and timing fully shielded wall packs to go off at ten P.M., except for one each on three sides of the building. The City of Marfa took this opportunity to include a notification clause in the lighting ordinance so that anyone applying for a building permit receives a copy of the code.

Not Left to Chance

Despite the Big Bend region's extensive history as a dark sky oasis, the reclamation of the nightscape was not left to chance. Civic changes occur as a direct result of advocates acting in a coordinated effort to create a culture within businesses, schools, legislative bodies, and members of the community. While the longtime presence of the McDonald Observatory undoubtedly jump-started the interest, it was collaborative citizen action that fostered the movement, and the citizens' legwork and perseverance that created the lighting changes and cultural awakening. Their efforts in the cities, developments, open areas, and parks have ensured a quality of life that is important to many Big Bend area residents, and are creating lasting protection for the ever-scarcer natural resource of a dark, starry night.

This triumph does not need to be extraordinary. Dark sky advocates are forming groups around the world and receiving assistance from conservationists, community activists, even forward-thinking developers and land-use experts. The sheer wonder inspired by a clear view of the cosmos crosses cultural and geographical boundaries, and, increasingly, serves to unite people toward a common goal.

Acknowledgments

This book would not have been possible without the contributions of many talented minds. The International Dark-Sky Association is honored and grateful to have received information, and material, in varying degrees, from the following people:

Dr. Mario Motta wrote most of the material in Chapter 2, "Disability Glare." Information in Chapter 3, "Effects on Human Health," was compiled by Dr. Steven Lockley, based upon evidence gathered in studies performed by Dr. Lockley and colleagues.

The discussion and case study of hatchling sea turtles in Chapter 4 was eloquently written by Dr. Blair Witherington of the Florida Wildlife Conservation Commission; apologies for any inadvertent mangling of an elegant and informative text.

IDA thanks Love Albrecht Howard, the author of Chapter 5 and a passionate and talented residential landscape designer, for sharing the principles of her trade so expressively.

Much of Chapter 6 was written by the late Timothy Crowe, a leader of the CPTED philosophy, which revolutionized urban surveillance methods across the United States. His generosity of time, knowledge, and spirit will never be forgotten.

The enlightening yet easy-to-follow discourse on commercial, industrial, and street lighting in Chapter 7 was authored by Jim Benya, proprietor of Benya Lighting Designs.

Suggestions for implementing public policy in Chapter 9 were generously provided by Brandi Smith.

Information and text on the tremendous victories in night sky conservation made by advocates in the Big Bend Region of Texas in Chapter 11 was provided by Gil Bartee, whose on-the-ground efforts in Alpine continue to protect its breathtaking night sky.

Sidebar authorship, acknowledged in the text, is again appreciated on this page. Many thanks to Paul Bogard, Susan Harder, and the Fatal Light Awareness Program's Irene Fedun for sharing their stories of darkness.

IDA wishes to thank numerous other people for their time and effort, and for their valuable insights that helped shape this book and broaden understanding of the issues surrounding light pollution. Special acknowledgement to Martin Morgan-Taylor, Connie Walker, Kevin Govender of the Southern African Large Telescope, and Kerri Robins, Cora-Lee Storvoldand Romeo Trastanetz of the City of Calgary.

Sincere thanks to the photographers featured in the book and who generously offered their images for publication.

Thanks to Dr. David Crawford and Dr. Tim Hunter, whose early action has improved the night vision of astronomers and stargazers worldwide, and to the countless unnamed (in this volume) but not unappreciated advocates, defenders, and guardians of the night sky. Keep looking up!

> *Is there not*
> *A tongue in every star that talks with man,*
> *And wooes him to be wise? nor wooes in vain;*
> *This dead of midnight is the noon of thought,*
> *And wisdom mounts her zenith with the stars.*
>
> — Anna Letitia Barbauld,
> "A Summer Evening's Meditation"

About the International Dark-Sky Association

Since 1988, the International Dark-Sky Association (IDA), has been rounding up stray light to save energy, safeguard ecosystems, and defend the view of the fading night sky. IDA is a 501(c)(3) nonprofit organization based in Tucson, Arizona, where the clear desert skies and effective lighting ordinances still present a compelling view of the stars. The IDA team educates individuals, scientists, and communities around the world about the consequences of poor lighting practices and offers affordable, ecologically forward solutions.

Light pollution is an intensely interdisciplinary topic, involving technology, astronomy, physics, biology, human health, ecology, energy and sustainability, politics, engineering, cultural history, and the arts. IDA's mission statement—"to preserve and protect the nighttime environment and our heritage of dark skies through environmentally responsible outdoor lighting"— acknowledges our commitment to a comprehensive approach in addressing the issues surrounding protection of the dark night sky.

To accomplish this multifaceted goal, IDA crafts effective initiatives (such as the Fixture Seal of Approval and International Dark Sky Places programs), pursues varied collaborations, and harnesses the enthusiasm of people who believe the night sky should not be transformed or forgotten. IDA has members in sixty-eight countries and a network that extends beyond the stars. Information sharing with ecologists, researchers and engineers provides insight on how advances in lighting technology can be applied to protect wildlife, save energy, and reduce carbon emissions. Our partnership with astronomers allows access to cutting-edge data on sky brightness. We work with large and small anti–light pollution groups such as the United Kingdom's Campaign for Dark Skies and Canada's Light Efficient Cities to promote control of artificial light worldwide, and share major projects with the National Park Service and the Illuminating Engineering Society.

In 2009, IDA opened the Office of Public Policy in Washington, DC, to promote research on lighting technology, educate federal officials on the resources saved by smart lighting policies, and work with government bodies to advance sound energy legislation. This office's outreach to the Department of Energy and the Environmental Protection Agency is laying the groundwork for policy change at the federal level.

Yet the backbone of IDA has always been our members. Many become talented volunteers who use IDA's resources to create changes in their own communities. IDA has fifty-eight volunteer-operated chapters in sixteen countries and hundreds of individual advocates. Their dedication and hard work is directly responsible for the global revolution in outdoor lighting practices. Whether the reason is amateur or professional astronomy, personal inspiration, outdoor activity, resource and energy conservation, or urban design, everyone has a reason to protect the natural night sky. Who has not been affected by one of the countless works of literature or art inspired by the sky above? Who has never heard a creation story based on the movements of planets and stars? The dark sky movement inspires people for all kinds of reasons. IDA pushes for systemic change while supporting individual advocacy to achieve change.

History

IDA is effective because it uses institutional knowledge to leverage individual interest, empowering the movement to flourish and achieve change. Two individuals formed IDA because they perceived a dire threat to a resource beloved to both of them. Twenty-four years ago, amateur astronomer Tim Hunter and professional astronomer David Crawford didn't realize that IDA would become the symbol of a new ethos: the dark sky movement. They just knew awareness about light pollution had to be raised or a scientific wonder and an ancient frontier would soon be invisible.

Crawford first experienced light pollution as an astronomer at the Kitt Peak National Observatory outside Tucson. From the time he started working there in 1959, he had watched Tucson change from a sleepy outpost to a burgeoning metropolis. As the population grew, so did sky glow. Crawford spent many years working on light pollution issues with local, national, and international organizations to address this growing problem on the largest scale possible.

Hunter was a serious amateur astronomer, even as child living in the suburbs of Chicago. He bought his first telescope, a four-inch reflector, in 1956. In 1985, he made a slightly larger investment and built the Grasslands Observatory in the sparsely developed desert fifty miles southwest of Tucson. Relentless development of southern Arizona made Hunter realize that the night sky was a fragile resource. Hunter began to notice lighting around Tucson. He was aghast when the University Medical Center, where he worked, installed orange low-pressure sodium lighting, and astounded when he was told that the lighting had been recommended by the astronomers at Kitt Peak.

Hunter discovered that professional astronomers prefer such lighting because its monochromatic spectrum is easy to avoid with high-powered telescopes. But the unshielded bulb created an unsightly light nuisance to amateur astronomers. Hunter knew that the Dark-Sky Office at Kitt Peak was headed by one David Crawford, internationally renowned as the project leader for the largest telescope at the observatory and for his work on stellar photometry. The two began to meet regularly.

They realized that light pollution is a relatively easy environmental problem to solve, but it must first be defined and recognized as a threat. Hunter had recently incorporated the Tucson Amateur Astronomy Association as a nonprofit organization and recommended that he and Crawford found a nonprofit to combat light pollution specifically. IDA was on its way.

Crawford expected it to be "an active international organization, wide in scope, breadth, and depth of topics. We wanted to show how all the issues related to dark skies were relevant to each other." He combined the new IDA goals into his professional work, and for many years devoted most of his waking hours to developing IDA into the organization it is today.

The organization has come a long way since Crawford and his wife, Mary, would produce the group's newsletter from an office in their home. Hunter proudly notes that "IDA has far outgrown its founders and surpassed our every hope for it."

David Crawford (left) and Tim Hunter faced an uphill battle convincing the world that light pollution was a problem that could be solved.

Outreach

IDA promotes one simple idea: Use only the light you need, only when you need it. Truly considering the question of how much light is actually necessary in modern society prompts examination of how we as humans perceive and interact with our environment. Such questions need to be asked at this time in our history, when the energy decisions of today are certain to have a profound effect on future generations. When people start to think critically about lighting, a transformation starts to occur. One becomes aware of how glare or light clutter detracts from ambiance. One questions how many natural resources are used to keep unnecessary or distracting lights on all night. One starts to become aware of opportunities for saving energy and restoring the nighttime environment to its natural state. At this point, the principles of dark sky–friendly lighting become innate, and it's easy to visualize the improvements brought by light used intentionally and sparingly, instead of by default.

IDA uses these sensible principles to develop tools that promote sustainable lighting practices. Brochures, posters, and presentations are available for free on the IDA website. News updates and information on scientific, educational, and conservation opportunities are available via email or Facebook. The IDA/IES Model Lighting Ordinance is intended to expedite adoption of lighting ordinances in North American cities. IDA's first white paper, "Visibility, Environ-

The International Dark-Sky Association protects the legacy of the night sky to inspire future generations. JANE KING

mental, and Astronomical Issues Associated with Blue-Rich White Outdoor Lighting," helped influence the direction of LED development. The IDSPlaces program and IDA's work with the National Park Service help assure that there will always be dark night skies to behold, and ways to share the night's hidden mysteries.

Positive Results

While light pollution continues to increase slowly, evidence shows that it is decreasing in parts of the world that have enacted effective legislation. A growing number of individuals are embracing the dark night after decades of attempting to banish it. Together, IDA, its chapters, affiliates, and individual volunteer efforts are reshaping the culture. Much like the growth of the "slow food" movement in response to the over-abundance of processed foods, the dark sky movement shows people that light for light's sake is not only undesirable, it is destructive to health and environment. Dark sky advocates will continue to seek a balance between our need to experience the night and our desire to control it.

This movement is growing in scale. Communities across the country are enacting lighting ordinances and the Model Lighting Ordinance will bolster widespread adoption of lighting controls. Recommended practices for outdoor lighting are changing for the better as studies show that less light is needed than previously thought. Cities are utilizing timers and dimmers to adjust lighting levels according to time of day. Italy and Slovenia have passed national ordinances limiting stray light. Greece has an extensive educational effort regarding light pollution in schools. The United Kingdom has released studies concluding that many urban areas are overlit. Australia has national standards for lighting, including the control of obtrusive lighting, with an outdoor lighting policy in Sydney. Other countries and locales remain active. Ongoing efforts in Japan, Malta, Spain, France, Germany, China, and Turkey ensure that the dark sky movement is felt around the globe.

IDA continues to grow in policy, practice, and followers. Together we are keeping a little light in the world–by taking it out of the sky.

Contributors

James R. Benya

James R. Benya (PE, FIES, FIALD, LC) is principal of Benya Lighting Design with offices in West Linn (greater Portland), Oregon. An internationally recognized lighting designer, educator, and writer, Jim has won more than 250 lighting design awards worldwide for his residential and commercial projects. He is a fellow of the International Association of Lighting Designers, a fellow of the Illuminating Engineering Society of North America, a registered professional engineer, and a member of the Institute of Electrical and Electronic Engineers, and is Lighting Certified by the National Council on Qualifications for the Lighting Professions. Jim currently serves as chairman of the Board of Fellows and member of the Technical Review Council of IESNA.

Paul Bogard

Paul Bogard is the editor of *Let There Be Night: Testimony on Behalf of the Dark* (U of Nevada P, 2008). His creative nonfiction has appeared in such places as *Creative Nonfiction, River Teeth,* and *The Gettysburg Review,* and he has written more than fifty articles for publications such as *Outside, Audubon, Backpacker,* and the *Albuquerque Journal.* Paul has a PhD in Literature and Environment at the University of Nevada, Reno, and now teaches at Wake Forest University in North Carolina.

Timothy Crowe

Timothy Crowe was a criminologist specializing in consulting and training services in law enforcement, crime prevention, juvenile delinquency and control, and major event law enforcement services planning. In addition to serving in state government and with several consulting firms, Mr. Crowe has provided security services for Republican and Democratic national conventions. He has served as the Director of the National Crime Prevention Institute at the University of Lousiville and created the NCPI's CPTED training program.

Tim's dedication to the theories behind Crime Prevention Through Environmental Design (CPTED) has helped revolutionize ideas in urban planning and crime fighting. His book *Crime Prevention Through Environmental Design, 2nd Edition,* contributed enormously to the growing popularity of this innovative approach to crime control. Mr. Crowe conducted CPTED in Schools training programs in Florida, North Carolina, Virginia, Kentucky, and the National School Resource Officer Program, as well as over 180 general CPTED courses. Timothy Crowe died on February 21, 2009, approximately seven weeks after the first draft of his article for this book was written.

Irene Fedun

Irene Fedun is a founding member of the Fatal Light Awareness Program and the editor of FLAP's biannual newsletter, *Touching Down.* She also edits the quarterly newsletter of the North American Native Plant Society, *The Blazing Star.* Irene envisions a world where humans live in delightful harmony and exquisite balance with the natural world.

Kevin Govender

Kevin Govender is manager of the Southern African Large Telescope Collateral Benefits Programme in South Africa. He is also Chair of the IYA2009 Cornerstone Project "Developing Astronomy Globally," formed to establish regional structures and networks for astronomy around the world in places that are not known for astronomy in order to stimulate social and economic benefits.

Susan Harder

Susan Harder is a resident of East Hampton, NY, and New York City. A former art dealer, she is a ten-year volunteer dark sky advocate who has delivered more than seventy-five public lectures to civic, environmental, and governmental groups. She is responsible for writing and supporting over a dozen outdoor lighting laws to help protect the night sky and reduce glare, light trespass, and excessive lighting. She is the recipient of an IDA Executive Director's award and a number of proclamations and citations from Long Island groups for her work to help communities address light pollution.

Love Howard

Love Albrecht Howard has owned her landscape design business since 1994. Her design work has been featured in national magazines and on the HGTV and DIY cable networks. She is a member of the Association of Professional Landscape Designers, and the Garden Writers Association of America, as well as a host of horticultural societies. She is the author of *So You Want to be a Garden Designer,* published by Timber Press, February, 2010.

Steven Lockley

Dr. Steven Lockley is an assistant professor of medicine at Harvard Medical School and an associate neuroscientist in the Division of Sleep Medicine, Department of Medicine, at Brigham and Women's Hospital. He is also an honorary associate professor of Sleep Medicine at the Clinical Sciences Research Institute, Warwick Medical School, UK, an adjunct associate professor at the School of Psychology and Psychiatry, Monash University, Australia, and a research associate in sleep and chronobiology for the Woolcock Institute of Medical Research in Australia. He has co-authored over thirty-five articles on the subjects of light, circadian phase, and fatigue.

Mario Motta

Mario Motta, MD, is president of the Massachusetts Medical Society, and a member of the American Medical Association's Council on Science and Public Health. He is a full-time cardiologist at North Shore Medical Center in Salem, Massachusetts. He is also an amateur astronomer, and built a homemade 32-inch telescope in an observatory attached to his home, where he observes and researches with the American Association of Variable Star Observers and other groups. He has been given the Las Cumbras award from the Astronomical Society of the Pacific and the Walter Scott Houston award from the Astronomical League. Dr Motta is current vice president of the AAVSO and an IDA Board member.

Brandi Smith

Brandi Smith is a doctoral student at Clemson University. A lifelong amateur astronomer, she conducts a self-crafted research program that concentrates on the role of night and light in the outdoor recreation experience—an extension of her 2007 Master of Public Administration degree thesis work at the University of Alabama at Birmingham (UAB). Her research seeks to further define and understand society's values, beliefs, and norms regarding night and use of light at night, enabling the development of more effective ways to educate the public about smart night lighting. Her master's work and led to the 2010 establishment of the Good Lighting Practices fellowship at Clemson, of which she is the first recipient.

Blair Witherington

Dr. Blair Witherington is a research scientist with the Florida Wildlife Commission's Fish and Wildlife Research Institute. He has a doctorate in zoology from the University of Florida. He has contributed numerous scientific articles and book chapters on sea turtle biology and conservation. His books include an edited volume on loggerhead sea turtles and popular books on Florida beaches, seashells, and sea turtles.

Index